# THE VIEW FROM SPACE

ENVIRONMENT AND SOCIETY

KIMBERLY K. SMITH, EDITOR

# The View From Space

## NASA's Evolving Struggle to Understand Our Home Planet

Richard B. Leshner and Thor Hogan

Published by the University Press of Kansas (Lawrence, Kansas 66045), which was organized by the Kansas Board of Regents and is operated and funded by Emporia State University, Fort Hays State University, Kansas State University, Pittsburg State University, the University of Kansas, and Wichita State University.

Library of Congress Cataloging-in-Publication Data

Names: Leshner, Richard B., author. | Hogan, Thor, author.
Title: The view from space : NASA's evolving struggle to understand our home planet / Richard B. Leshner and Thor Hogan.
Description: Lawrence : University Press of Kansas, [2019] | Series: Environment and society | Includes bibliographical references and index.
Identifiers: LCCN 2019006959
ISBN 9780700628315 (cloth : alk. paper)
ISBN 9780700628322 (paperback : alk. paper)
ISBN 9780700628339 (ebook)
Subjects: LCSH: Earth Observing System (Program) | Earth sciences—Remote sensing. | Earth sciences—Observations. | Artificial satellites in remote sensing. | Artificial satellites in earth sciences.
Classification: LCC QE33.2.R4 L46 2019 | DDC 550.28/4—dc23
LC record available at https://lccn.loc.gov/.

British Library Cataloguing-in-Publication Data is available.

Printed in the United States of America

10 9 8 7 6 5 4 3 2 1

The paper used in this publication is recycled and contains 30 percent postconsumer waste. It is acid free and meets the minimum requirements of the American National Standard for Permanence of Paper for Printed Library Materials Z39.48–1992.

# Contents

*Acknowledgments vii*

*List of Acronyms and Abbreviations ix*

Introduction 1

1. A Global View of the Earth 29

2. Evolution of an Idea 51

3. A Long Road to a New Mission 77

4. Rearguard Action to Save EOS 119

5. Last Hard Push to Secure EOS 149

Conclusion 179

*Notes 207*

*Index 235*

# Acknowledgments

Our many thanks to the entire team at the University Press of Kansas, who have been incredible to work with through this entire process. We are hugely grateful to Kim Hogeland for seeing the potential in our initial manuscript and believing it would be a good fit for the press. We are obliged to Kimberly Smith for her confidence that the book would be a suitable addition to the growing Environment and Society collection. We are appreciative of all the work that Kelly Chrisman Jacques did during the production phase of the project. Finally, thanks so much to Michael Kehoe and all the other folks in the Marketing Department for getting the book ready for publication.

I am especially indebted to my graduate school advisors: many thanks to Ray Williamson, John Logsdon, Steve Balla, and Joe Cordes for their support and direction. This project would not have been possible without the matchless assistance I received from the entire team at the NASA History Office.

I would like to express my gratitude and many thanks to my grandfather for imbuing me with an interest in politics and history; my uncle, Joe, for his continued encouragement and for helping me get over the finish line; my wife and daughter for their continued inspiration; and, my mom, Susan, for teaching me how to work hard and strive to better myself—this work is a testament to her superlative parenting and unconditional love.
—Richard B. Leshner

This book would not have been possible without support from Earlham College, and I would like to thank Greg Mahler in particular for championing scholarly research at the college. I am especially indebted to several students who have engaged in student-faculty collaborative research projects with me over the past few years. David Schutt provided invaluable assistance in helping shape the theoretical discussion found in the conclusion of the manuscript. Eli Richman, Gerald Sowah, Cephas Toninga, and Scott Slucher drafted highly useful notes for studies that evaluated the impact of MTPE/EOS and provided helpful overviews of the accomplishments of every earth science satellite launched by NASA since 1998. Finally, I would like to thank the dedicated archivists at the George H. W. Bush Presidential Library at Texas A&M.

My love and appreciation goes out to Kate and Sam, who, as always, have put up with me during my work on this project.

—Thor Hogan

# Acronyms and Abbreviations

| | |
|---|---|
| ABMA | Army Ballistic Missile Agency |
| AgRISTARS | Agriculture and Resource Inventory Surveys through Aerospace Remote Sensing |
| AIRS | Atmospheric Infrared Sounder |
| ARPA | Advanced Research Projects Agency |
| ASTER | Advanced Spaceborne Thermal Emission and Reflection Radiometer |
| ATS | Applications Technology Satellite |
| AVCS | Advanced Vidicon Camera System |
| BOB | Bureau of the Budget |
| Caltech | California Institute of Technology |
| CAP | Comparative Agendas Project |
| CBO | Congressional Budget Office |
| CES | Committee on Earth Sciences |
| CGC | Committee on Global Change |
| CIAP | Climatic Impact Assessment Program |
| DARPA | Defense Advanced Research Projects Agency |
| DMSP | Defense Meteorological Satellite Program |
| DOD | Department of Defense |
| DOE | Department of Energy |
| DOI | Department of the Interior |
| DOT | Department of Transportation |
| EOS | Earth Observing System |
| EOSDIS | EOS Data and Information System |
| EPA | Environmental Protection Agency |
| ERBE | Earth Radiation Budget Experiment |
| EROS | Earth Resources Observation Satellite |
| ESA | European Space Agency |
| ESSC | Earth System Science Committee |
| FAA | Federal Aviation Administration |

| | |
|---|---|
| GAO | General Accounting Office |
| GLOBE | Global Learning and Observations to Benefit the Environment |
| GOES | Geostationary Operational Environmental Satellite |
| GOJ | government of Japan |
| GRM | Geospatial Research Mission |
| Goddard | Goddard Space Flight Center |
| HRIR | High-Resolution Infrared Radiometer |
| HIRIS | High-Resolution Imaging Spectrometer |
| ICBM | intercontinental ballistic missile |
| IRBM | intermediate-range ballistic missile |
| IGY | International Geophysical Year |
| IPCC | Intergovernmental Panel on Climate Change |
| IRIS | Infrared Interferometer Spectrometer |
| ITOS | Improved TIROS Operational Satellite |
| JPL | Jet Propulsion Laboratory |
| JSC | Johnson Space Center |
| LACIE | Large Area Crop Inventory Experiment |
| LIMS | Limb Infrared Monitor of the Stratosphere |
| MRIR | Medium-Resolution Infrared Radiometer |
| MSS | multispectral scanner |
| MTPE | Mission to Planet Earth |
| NACA | National Advisory Committee on Aeronautics |
| NASA | National Aeronautics and Space Administration |
| NEMS | Nimbus E Microwave Spectrometer |
| NOAA | National Oceanographic and Atmospheric Administration |
| NOMSS | National Operational Meteorological Satellite System |
| NRC | National Research Council |
| NSC | National Security Council |
| NSF | National Science Foundation |
| OMB | Office of Management and Budget |
| OMSS | Office of Manned Space Science |
| OPEN | Origins of Plasmas in Earth's Neighborhood |
| OSIP | Operational Satellite Improvement Program |
| OSSA | Office of Space Science and Applications |
| OSTP | Office of Science and Technology Policy |
| POES | Polar Orbiting Environmental Satellite |

| | |
|---|---|
| POP | Polar Orbiting Platform |
| RCA | Radio Corporation of America |
| RFP | request for proposal |
| SAIC | Science Applications International Corporation |
| SAMOS | Satellite Military Observation System |
| SAMS | Stratospheric and Mesospheric Sounder |
| SCAMS | Scanning Microwave Spectrometer |
| SDO | Solar Dynamics Observatory |
| SEI | Space Exploration Initiative |
| SIRS | Satellite Infrared Spectrometer |
| SMS | Synchronous Meteorological Satellite |
| SSCC | Spin-Scan Cloud Camera |
| SST | supersonic transport |
| STS | Space Transportation System |
| TDRSS | Tracking and Data Relay Satellite System |
| TIROS | Television Infrared Observing Satellite |
| TOPEX | Ocean Topography Experiment |
| TOS | TIROS Operational System |
| UARP | Upper Atmosphere Research Program |
| UARS | Upper Atmosphere Research Satellite |
| USAF | U.S. Air Force |
| USDA | U.S. Department of Agriculture |
| USGCRP | U.S. Global Change Research Program |
| USGS | U.S. Geological Survey |
| VOIR | Venus Orbiting Imaging Radars |

# Introduction

For nearly two decades, the National Aeronautics and Space Administration (NASA) has been directing one of the most important Big Science programs in human history. It fought for many years to gain approval for an expansive program in earth science research, but in recent years this battle has proved well worth fighting as various space agency satellites have become critical contributors to the American government's overarching objective of better understanding the evolving climate crisis—which makes it far easier to design and adopt policy interventions aimed at mitigating and adapting to what many believe is the most important problem facing civilization. Despite the overwhelming accomplishments of this undertaking, however, the entire mission is continually threatened by changes within the government. As a result, the time seems opportune for a book that investigates how and why earth science research became a NASA priority. Our hope is that this book will provide scientists, activists, and policymakers seeking to preserve these critical research projects with a tool to help them make their case to the American people.

While human spaceflight and space science have long been seen as the primary missions of NASA, the space agency's earth science research program may ultimately prove to have been far more important. In 1990, after nearly a decade of debate, NASA received its first budget appropriation for the development of the Earth Observing System (EOS). EOS was the technological cornerstone of a new research initiative, Mission to Planet Earth (MTPE), which aimed to use satellites to study the planet's environment from space. Part of the governmentwide US Global Change Research Program (USGCRP), an enterprise established under President George H. W. Bush, the goal of MTPE was to better understand fundamental earth processes such as weather and climate. At the time, the projected budget for MTPE was approximately $17 billion over ten years. This outlay would have enabled the launch of numerous platforms, including two large polar-orbiting EOS satellites.

By this time, NASA already had a long history of developing satellites capable of viewing the earth. During its early years, the space agency had developed the civilian meteorological satellites that were operated by the U.S. Weather Bureau. These satellites were followed by Landsat, which gave birth to civilian remote sensing from space. The significantly larger budgetary resources planned and allocated for MTPE, however, distinguished it from these past efforts and suggested that global change science (also known as earth system science) had become an important new mission for NASA. The space agency had begun developing this new initiative in the early 1980s, when a task force of experts explored the utility of satellites monitoring the environment. Based on the group's positive findings, NASA's leadership team began promoting an international project to use space assets to monitor and assess natural and human impacts on the habitability of the earth. While the underlying rationale for such a program was widely embraced, the agency faced domestic and international criticism for tying the program too closely to its own goals. As a result, the idea initially received little support. Nonetheless, the agency continued to develop mission concepts and lobby for the program as it waited for an opportunity to win support for the initiative from both the White House and Congress. This opportunity came with the election of President Bush, who provided key support for the program.

Following its adoption, MTPE/EOS began to undergo immediate and significant changes. In 1991, the program was forced to reduce its projected budget from $17 billion to $11 billion as the result of a government-wide effort to reduce federal spending in the face of growing deficits and a weakening economy. This reduction led to major satellite redesigns and modification of the initiative's overall goals. This initial *restructuring* effort was followed by three more redesign efforts, each following a budget cut. These continual changes were the result of several factors: efforts by the Clinton administration and Congress to reduce federal spending and reduce the NASA budget, changes in the technical rationale for EOS and its dependence on large spacecraft, new budget priorities at NASA, and changing attitudes about the priorities of MTPE/EOS within the scientific community. Regardless, in 1999, after nearly two decades of discussion and development, the program began returning scientific data when the Terra platform was launched and became operational. Today, NASA's earth science program includes nearly two dozen satellites that are providing integrated measurements of planetary processes. As concerns grow regarding the likely impacts of global climate change, the scientific data provided by these platforms continually informs the policy debate over appropriate government interventions while at the same time teaching us more about the nature of our home planet.

## Evolution of the American Space Program

It is impossible to fully understand the evolution of MTPE/EOS without first appreciating the historical development of the American space program. This record provides critical context for comprehending how and why the earth science initiative came to be adopted. While chapter 2 will provide a detailed overview of early remote sensing programs, in this section we will briefly consider how these efforts fit into the maturation of the overarching civilian space program.

In many ways, the roots of the program stretch back many centuries to when astronomers, writers, and filmmakers first defined humanity's dreams of spaceflight. In 1634, the German astronomer Johannes Kepler wrote *Somnium*, a remarkably prescient story about a voyage to the moon that was also a treatise on lunar astronomy. Although it went unrecognized for centuries, it ultimately became known as one of the most important books in the history of science. More than two centuries later, in 1865, Jules Verne wrote *From Earth to the Moon*, the tale of three members of a post–Civil War gun club who build an enormous sky-facing cannon to launch a spaceship to the moon. Four years later, Edward Hale published *The Brick Moon*, which was the first proposal for an Earth-orbiting satellite. A couple of decades afterward, the American astronomer Percival Lowell began writing about his observations of Mars, popularizing the idea that the red planet was home to a dying race that had built a vast network of canals to transport limited water supplies around their world. At the turn of the twentieth century, H. G. Wells wrote *War of the Worlds* and *The First Men on the Moon*; the former was history's most famous alien invasion story, while the latter proved to be a prophetic description of how humans might reach the lunar surface. Finally, in 1902, the legendary French director George Méliès produced *Le Voyage dans la Lune*, the silver screen's first science fiction film and a massive popular success on both sides of the Atlantic Ocean.[1]

These astronomers, writers, and filmmakers fired the imaginations of the true visionaries, who would turn fantasy into reality. During the early twentieth century, innovation in spaceflight largely flowed from the work of five technologists. In 1903, Russian physicist Konstantin Tsiolkovsky published "Exploration of Space with Rocket Devices," the seminal work that laid out the scientific argument regarding how a chemical rocket could allow humans to break gravity's hold and orbit Earth. Tsiolkovsky's proposed missile would utilize a mixture of liquid oxygen and liquid hydrogen, which upon combustion would expand and be thrust out of an exhaust hole at high velocity. The following decade, the German physicist and engineer Hermann Oberth began a research project examining this reaction principle. Before World War I, he proposed the construction of liquid-fueled rockets to the German military, but his idea was rejected. In 1923, he published his influential book *The Rocket into Planetary Space.* As Roger Launius and Ray Williamson contend, his manuscript was

> a thorough discussion of almost every phase of rocket travel. [Oberth] posited that a rocket could travel in the void of space and that it could move faster than the velocity of its own exhaust gases. He noted that with the proper velocity a rocket could launch a payload into orbit around the Earth, and to accomplish this goal he reviewed several propellant mixtures to increase speed. He also designed a rocket that he believed had the capability to reach the upper atmosphere by using a combination of alcohol and hydrogen as fuel.[2]

Oberth went on to found the German Society for Space Travel, which cultivated an entire generation of rocket scientists. While Tsiolkovsky's and Oberth's work was largely theoretical in nature, their ideas set the foundation for much of the rocketry research conducted by Germany during World War II.[3]

On the other side of the world, working independently of his European counterparts, American Robert Hutchings Goddard reached similar conclusions regarding the potential for rockets to boost humans into Earth orbit and beyond. Unlike the Russian and German theorists, however, Goddard pursued a course of not just thinking about rockets but actually building and testing them. Born twenty-five years after Tsiolkovsky,

Goddard submitted speculative articles regarding space navigation and multistage rockets to *Popular Science News* as early as 1902, while he was still in high school. He received a doctorate in physics from Clark University in Massachusetts, where he began his initial calculations relating to liquid-fueled rocket engines. In 1917, Goddard received a grant from the Smithsonian Institution to conduct research on high-altitude rockets. Two years later, he published the first of his great works, entitled "A Method of Reaching Extreme Altitudes." The paper explained how a reusable two-stage solid-propellant rocket could be used to conduct experiments in the upper atmosphere. He also suggested, in a small section of the manuscript, that liquid-fuel propulsion might be capable of taking humans to the moon. Although the manuscript is now considered a great scholarly work, it was largely misunderstood at the time and led the media to label Goddard "the Moon man" due to his speculation about journeys to the moon. This unexpected reaction from the press horrified the American physicist, leading him to conduct most of his cutting-edge research in complete seclusion from the larger scientific community. In 1926, Goddard officially initiated the age of modern rocketry when he successfully launched a 10-foot-long rocket on a 2.5-second flight; it reached an altitude of 41 feet and flew a distance of 184 feet. He continued to refine his designs over the subsequent two decades, eventually launching his L-13 rocket to an altitude of 1.7 miles. Although his contributions were largely overlooked and even ridiculed by his contemporaries, later generations recognized the brilliance of his work, and he is now known as the father of modern rocketry.[4]

One of the engineers inspired by Herman Oberth's theoretical work in rocketry was a German engineer named Wernher von Braun. In 1932, this young aristocrat, who was an early member of Oberth's Society for Space Travel, was chosen to lead the German military's rocket artillery unit. Over the next decade, von Braun assembled a team of scientists and engineers that conducted ongoing research at the Peenemunde facility on the Baltic coast. The rocket team ultimately developed the V-2. This so-called *Vengeance Weapon* was designed, tested, and built as an operational liquid-fuel rocket intended to frighten Allied troops and civilian populations. Over 5,000 were constructed at the 900,000-square-foot underground production plant in Nordhausen. The 50-foot-tall vehicle could propel 2,310 pounds of explosives over 190 miles, flying up to 60 miles high at a peak speed of 3,500 miles per hour. This represented an amazing advancement in rocket technology. By the end of the war, approximately 3,225 V-2s had been fired, primarily at Antwerp and London. As the war came to an end, von Braun and more than one hundred members of his rocket team traveled across Germany and surrendered to advancing American forces. The Americans rushed to Peenemunde and Nordhausen and seized all of the remaining rockets. This collection of scientists and flight materials formed the core of early ballistic missile research conducted in the United States.[5]

Born three years after Tsiolkovsky published "Exploration of Space with Rocket Devices," Sergei Korolev was the chief designer of the Soviet launch vehicle that opened the space age. Korolev was trained in aeronautical engineering at the Kiev Polytechnic Institute and founded the Russian rocket society in the early 1930s. He spent the majority of World War II in a sharashkas (prison design bureau) working with senior aircraft engineer Sergei Tupolev. In 1946, Korolev was released from prison and appointed chief designer for the development of long-range rockets. During the next decade, his work focused primarily on designing, testing, and building a series of ballistic missiles for the Soviet military. In 1954, he was tasked with constructing the nation's first intercontinental ballistic missile (ICBM). He took advantage of the opportunity to request that his design bureau also be allowed to study the possible use of this technology for spaceflight, with the eventual goal of launching an artificial satellite. His request was accepted, and a modest research program was initiated.[6]

Events outside the Soviet Union played an important role in the rapid evolution of Korolev's research. In the early 1950s, a group of American and European scientists led by James van Allen had formulated the idea for the International Geophysical Year (IGY). The IGY was designed as an international effort to examine the upper atmosphere and outer space utilizing sounding rockets, balloons, and ground observations. The period from 1 July 1957 to 31 December 1958 was predicted to be a time of unusual solar activity. In 1955, the year after Korolev began investigating options for using the R-7 ICBM to launch a satellite, the United States announced that it was planning to do just that during the IGY. This revelation took the Soviets by surprise and led to increased funding for the nation's space efforts. Korolev convinced leading Soviet officials that launching an artificial satellite, thus demonstrating the nation's scientific and technological prowess to the world, would have dramatic political repercussions. Over the next year, Korolev finalized the changes to the R-7 required to use it for a satellite launch. At the same time, Soviet scientist Mikhail Klavdiyevich

Tikhonravov was leading the effort to construct a relatively complex satellite named Object D. Those plans were subsequently scaled down because it was not clear that the R-7 could provide the thrust necessary to send the heavy satellite into orbit, and a new design for a smaller and simpler satellite was approved. In 1957, the early launch attempts of the new R-7 configurations failed, calling into question whether Korolev could send the satellite into orbit during the IGY. A successful flight test in the early fall, however, paved the way for a full-on attempt to launch the satellite. On October 4, the five-engine R-7 rocketed into the night sky and successfully placed the world's first artificial satellite, dubbed *Sputnik*, into an elliptical orbit around Earth. Historian Asif A. Siddiqi writes that "the political and cultural shock bequeathed by *Sputnik* set events in motion that eventually gave rise to Apollo, perhaps the central artifact of the so-called 'space race' of the Cold War." In turn, the space race served to drive a massive American investment in research in an effort to attain world leadership in outer space.[7]

While Sergei Korolev was developing ICBMs in the Soviet Union, there were ongoing efforts within the United States to do the same thing. Concentrated missile research began with testing and experimentation of the V-2 rockets that had been confiscated in Germany after the war. Throughout the subsequent development of a homegrown American capability, a strong connection was forged between the development of weapon-carrying ballistic missiles and Earth-to-orbit launch vehicles. The V-2 test program was used not only to achieve technological parity with the defeated Germans but also to promote spaceflight for scientific reasons. This was the basis for the development of the early American sounding rockets that were used by James van Allen to conduct investigations of the planet's upper atmosphere. From 1946 to 1951, the U.S. Army conducted nearly seventy test flights of original and modified V-2s at the White Sands test facility in New Mexico. That research provided groundbreaking new data related to controlling the ignition and separation of multistage rockets and also contributed to advanced rocket engine development. Pioneering work begun during that time by North American Aviation's Rocketdyne division led to the development of the XLR83, which could produce an astonishing 415,000 pounds of thrust (this engine was a precursor of the F-1, the first stage rockets on the Saturn V launch vehicle). During the same period, the U.S. Naval Research Laboratory was also conducting sounding rocket research using the Viking launch vehicle.[8]

As the nation moved into the 1950s, Cold War tensions grew during the Korean War, and nuclear weapons testing expanded, the American military established a number of different ballistic missile programs. One that would prove particularly important for the nation's entry into the space age was the Redstone—a highly accurate, liquid-propelled, surface-to-surface missile capable of transporting nuclear warheads up to 200 miles. Wernher von Braun and his German rocket team developed the Redstone at the Army Ordnance Rocket Center in Huntsville, Alabama. The Jupiter program, a concurrent joint project undertaken by the army and navy designed to construct a missile capable of being launched from land or sea, was an expansion of the Naval Research Laboratory's work that led to the development of the Viking sounding rocket. Two versions of the Jupiter were built, both using the Redstone as a first stage. The Jupiter-A was an intermediate-range ballistic missile (IRBM) designed to carry a nuclear warhead, while the Jupiter-C was intended to test ballistic reentry technologies. The Jet Propulsion Laboratory (JPL), a research institute at the California Institute of Technology, provided the second and third stages of the Jupiter-C. This combination was successfully flown on 8 August 1957 (two months before *Sputnik*) to a height of 600 miles and downrange 1,000 miles.[9]

When the United States announced that it would launch an artificial satellite during the IGY, a Department of Defense (DOD) committee was formed to select the vehicle that would boost the satellite into orbit. The committee had three different rockets to choose from: the air force's Atlas ICBM, which was still in development; the army's Jupiter-C; and an unnamed navy vehicle utilizing a Viking first stage, the Aerobee sounding rocket as a second stage, and a new solid-fuel third stage. Believing that the third proposal had more merit because it was based on a framework of existing rockets, the committee selected what became known as Project Vanguard. There were considerable technical problems, however, in getting these three rocket stages to work together. The first launch of the vehicle, which had been announced by President Dwight Eisenhower shortly after the launch of *Sputnik*, was a dramatic failure. Launius and Williamson argue that the failure of Vanguard "set a tone that carried through the early days of the U.S. space program. Not only had the Soviets been first in space, but the [failure of Vanguard] heightened the perception that U.S. engineers were space bunglers and stiffened U.S. resolve to best the Soviets."[10] The very public failure of Vanguard opened the door for von

Braun's Army Ballistic Missile Agency (ABMA) and the Jupiter-C. On 31 January 1958, a modified Jupiter-C with a fourth stage (this configuration was named the Juno 1) boosted the first American satellite into orbit. The *Explorer 1* was tiny, weighing only 18 pounds, but carried instrumentation that was used to discover the Van Allen radiation belts—a much greater accomplishment than that of *Sputnik*. The successful launch of America's first satellite set the stage for the nation's efforts to compete with the Soviet Union in the arena of outer space.[11]

The American public's response to the launch of *Sputnik* was earsplitting. The public outcry was largely a reaction to the perceived threat to national security posed by Earth-orbiting satellites. Historian Walter McDougall argues that this was the greatest challenge of Eisenhower's tenure because his administration was charged with jeopardizing America's position in the world through its complacency. In the aftermath of the Soviet space achievement, the rest of the world began to look at both countries in a different light. The communists appeared to be gaining technological superiority as the capitalists appeared to be waning.[12]

The launch of *Sputnik* and the American public's reaction to that event produced a political tidal wave that led to one of the most important battles of the Cold War. As McDougall notes, the American response to the launch of *Sputnik* changed the

> fundamental relationship between the government and new technology . . . as never before in history. No longer did state and society react to new tools and methods, adjusting, regulating, or encouraging their spontaneous development. Rather, states took upon themselves the primary responsibility for generating new technology. This has meant that to the extent revolutionary technologies have profound second-order consequences in the domestic life of societies, by forcing new technologies, all governments have become revolutionary.[13]

The combination of political circumstance and the state of technology produced a unique historical situation that resulted in some of the most striking technological achievements in human history. It also resulted in a dramatic shift in the American government's attitude toward the importance of federal investment in civilian aerospace research and set the stage for staggering increases in funding for space exploration.

Eisenhower attempted to calm the public by telling Americans that the nation's space program was in good shape. In a White House press conference, he stated that American ICBM and IRBM programs continued to be a top national priority and that the launch of *Sputnik* did not represent an immediate threat to national security. By all indications, Eisenhower took the position that America should not react in haste to the Soviet launch. He feared that if Congress initiated a crash satellite program, it would lead to overspending that would endanger the economy. Thus, he stated that the United States would continue its programs as scheduled and would not allocate emergency funds. Eisenhower's low-key response did little to allay public fear. On the contrary, many Americans became more worried because they believed that the Eisenhower administration was not taking the issue seriously. In a poll conducted in early 1958, 66 percent of those surveyed said they believed that the Soviets were beating the United States in the Cold War.[14]

Although the launch of *Explorer 1* considerably reduced the public turmoil, the proposal for a national space agency first surfaced in its immediate aftermath. On 6 February 1958, the Senate established the Senate Committee on Outer Space, chaired by Majority Leader Lyndon Johnson, to establish a national space policy and examine the need to create a space agency. Eisenhower was initially opposed to the idea of a civilian space agency because he believed that any new organization should be under military leadership. Newly appointed Science Advisor James Killian and Vice President Richard Nixon convinced the president that a civilian agency would better suit the peaceful character that they believed was imperative for the American space program. As a result, Eisenhower assigned the Presidential Science Advisory Committee and Killian to design a civilian space agency with a largely scientific agenda. The committee outlined a program that could include the development of weather and communications satellites, lunar and planetary probes, and a human mission to the moon.[15]

On 2 April, Eisenhower sent a special message to Congress about space science and exploration. The communication stated that the advancement of space technology should be pursued for four major reasons: the compelling urge to explore the unknown, the need to take full advantage of the military potential of space, the effect on national prestige of accomplishment in space science and exploration, and opportunities for scientific observation and experimentation. The letter recommended that "aeronautical and space science activities sponsored by the United States be conducted under the direction of a civilian agency. . . . I am, therefore, recommending that the responsibility for administering the civilian space science and exploration program be lodged in a new Aeronautics and Space Agency."[16]

In May 1958, the Soviet launch of *Sputnik III* (weighing over 3,000 pounds) proved once again that America was trailing badly in rocket technology. This produced an impetus for speeding the establishment of an organization to coordinate the nation's space efforts. Lyndon Johnson wrote and sponsored the National Aeronautics and Space Act, which was easily passed by Congress early that summer and was signed by Eisenhower on 29 July 1958. The act officially established NASA and assigned all nonmilitary rocket programs to the new organization. The new administration inherited facilities across the country, including all of the research centers and test sites of the National Advisory Committee on Aeronautics (NACA), the ABMA, and JPL. Eisenhower chose Keith Glennan as the first administrator of the space agency. Under his leadership and with support from the White House, NASA received a budget of $240 million for fiscal year 1959—almost double the peak amount received for previous civilian programs.[17]

In the immediate aftermath of the creation of the program, its top priority was to launch the first human into space. For this effort, named Project Mercury, NASA had to grapple with new and complex systems engineering problems that had never previously been addressed. A key issue was managing risk, balancing program costs against the desire to fulfill the mission objectives. A large part of this process involved integrating sufficient redundancies into existing launch vehicles so that single-point failures that could lead to the loss of crew members were completely understood and eliminated wherever possible. All told, NASA would spend $392 million on Project Mercury from 1959 to 1963. The primary objective of the program was "to achieve at the earliest practical date orbital flight and successful recovery of a manned satellite, and to investigate human capabilities in this environment."[18]

Senator John Kennedy was not heavily engaged in space policy even after the Soviet launch of *Sputnik*. As his 1960 presidential campaign took shape, however, he began to investigate the issue more closely. His campaign advisers believed that the perception that the Eisenhower administration had not adequately prioritized the ballistic missile program could be used against Richard Nixon during the upcoming election. Therefore, an important part of Kennedy's campaign strategy was to criticize the Republican White House for failing to match the Soviets in the space race and to suggest that this failure had jeopardized national security. To remedy the problem, Kennedy endorsed increased research funding to accelerate the missile and space programs—including expanded spending for basic research in space and rocket technology.[19]

After winning one of the narrowest electoral victories in American history, Kennedy decided to assign responsibility for the space program to Vice President–elect Lyndon Johnson. The new vice president was highly qualified for the position and knew more about the issue than anyone else in the administration. The new president had requested several reviews of national space efforts, but these had offered mixed recommendations regarding the appropriate future course for the program. During the early months of his tenure, space policy took a backseat to more pressing matters. In April 1961, however, when Soviet cosmonaut Yuri Gagarin became the first human in space and the Bay of Pigs invasion failed, Kennedy became more heavily involved. To address these two political disasters, the youthful president decided that a program that could shift public attention toward a bold new initiative was needed.[20]

Over the course of the next several weeks, a rapid policy review was conducted to determine whether there was any "space program which promises dramatic results in which we could win." This review revealed that a crewed lunar mission, rather than a space station or circumlunar mission, was the policy option with the best chance of beating the Soviets. On 25 May, a month after the successful launch of Astronaut Alan Sheppard in a Mercury capsule, President Kennedy addressed Congress in a speech titled "Urgent National Needs." The major themes of the speech were economic and social progress at home and abroad, national security, civil defense, and nuclear disarmament. The president concluded his remarks with an explanation of his vision for the American space program. He famously said,

> Now it is time to take longer strides—time for a great new American enterprise—time for this Nation to take a clearly leading role in space achievement which in many ways may hold the key to our future on earth. . . . I therefore ask the Congress, above and beyond the increases I have earlier requested for space activities, to provide the funds which

are needed to meet the following national goals. . . . I believe that this Nation should commit itself to achieving the goal, before this decade is out, of landing a man on the moon and returning him safely to earth.

Kennedy's requests received overwhelming support in the Congress. In addition, the public reaction to the speech was very enthusiastic. By the middle of the summer, both chambers of Congress had endorsed Project Apollo, which would ultimately bring NASA's share of the federal budget to 4.5 percent at more than $35 billion annually (in fiscal year (FY) 2018 dollars). Ultimately, the program was highly successful. Starting in 1969, and for a period of just over three years thereafter, NASA successfully landed twelve humans on the moon.[21]

Throughout the Apollo era and during the subsequent half century, the space agency conducted a large amount of advanced research that had no connection to the human spaceflight program. The civilian space program was largely responsible for fostering the fundamental technologies that fueled the telecommunications revolution. The organization conducted revolutionary space science research that fundamentally changed how we understand our solar system and our universe. All of these various endeavors shaped the ongoing commitment to the national space program.

In 1945, a British Royal Air Force officer named Arthur C. Clarke wrote an article that described the use of crewed orbital satellites to transmit television programs around the globe, for which he is commonly referred to as the father of satellite communications. John Pierce at Bell Laboratories carried out the first serious work to advance this idea. His research found that a satellite capable of transmitting one thousand calls at the same time was theoretically possible. In 1962, these efforts led to the launch of the experimental *Telstar* satellite for AT&T. While Bell Labs continued its applied work, NASA was also proceeding with an aggressive research program to investigate the required technologies for satellite communications. This led to a series of experimental satellites, including *Echo* (the first artificial satellite that actually relayed a real-time voice message from Earth to orbit and back), *Relay*, and *Syncom*. The government-backed Communications Satellite Corporation adopted technologies from these programs for its *Early Bird* satellites, which formed the backbone of America's early satellite telecommunications network.[22]

The origins of space science are rooted in early balloon research conducted in the United States shortly after the conclusion of World War II. With the advancement of rocket technology in the postwar years, however, researchers increasingly turned to sounding rockets to perform investigations of the upper atmosphere. As discussed above, *Explorer 1* was designed to conduct space science research and led to the discovery of the Van Allen radiation belts. The early years of NASA's space science program were often chaotic, with various internal and external players vying to prioritize the agency's space science missions. The creation of the NASA Office of Space Science was an important organizational stepping-stone, providing the agency with a strong mandate for space science and two field centers (the Goddard Space Flight Center and JPL) to conduct related activities.[23]

During the Apollo era, the majority of NASA's space science activities revolved around lunar exploration and sending planetary probes to Venus and Mars. In March 1959, the *Pioneer 4* spacecraft (which returned important radiation data and acted as a valuable tracking exercise) successfully passed within 37,300 miles of the moon. From January 1964 to March 1965, four *Ranger* spacecraft conducted significant lunar research and returned more than seventeen thousand photographs of the lunar surface. In May 1966, *Surveyor 1* performed the first American soft landing on the moon. Additional missions in the *Surveyor* series achieved additional milestones, including a chemical analysis of the lunar soil using a robot arm (in total, the five landers also returned nearly ninety thousand images of the lunar surface). During the same period, the *Lunar Orbiter* program conducted five orbital missions around the moon designed to produce high-quality photographs of the surface. Over the course of the two-year project, these spacecraft produced amazing lunar maps and photographed twenty potential landing sites for the Apollo program.[24]

From August 1962 to March 1975, at the same time that NASA was conducting intense lunar research in support of the Apollo program, seven *Mariner* probes conducted initial investigations of Earth's three planetary neighbors in the inner solar system. These missions racked up an astounding track record, including measuring the solar wind, returning close-up photographs of planetary surfaces, analyzing planetary atmospheres, and pioneering gravity assist as a method for navigating the solar system. The highly ambitious Viking missions, designed to place two landers on the Martian surface, followed these initial planetary explorations. In 1976, *Viking 1* touched down on the western slope of Chryse Planitia and *Viking 2* settled down at Utopia Planitia. These two missions returned incredible photographs of Mars and performed three biology experiments designed to look for possible signs of life (while these experiments discovered mysterious chemical activity in the Martian soil, they provided no clear evidence of the presence of living microorganisms in soil near the landing sites). *Voyager* was the final great space science program of the 1970s, launching two spacecraft (*Voyager 1* and *Voyager 2*) during the second half of 1977. Between them, the two explored all the giant planets of the outer solar system (Jupiter, Saturn, Uranus and Neptune), nearly fifty of their moons, and the unique system of rings and magnetic fields possessed by those planets. Both missions are still operational and are conducting invaluable interplanetary research in the region of space where the sun's influence ends and deep space begins.[25]

By the 1980s, the success and expense of the *Viking* and *Voyager* missions, combined with increasingly tight budgets, led NASA to reduce its commitment to planetary exploration. It was not until 1989, more than a decade after the launch of *Voyager 2*, that the space agency launched the *Magellan* and *Galileo* spacecraft. *Magellan*, which arrived in orbit around Venus in 1990, provided earthbound researchers with invaluable radar maps of the Venusian surface and atmospheric data during its final plunge into the planet. Likewise, the *Galileo* probe, which arrived in orbit around Jupiter in 1995, compiled an impressive record of achievement in its exploration of the Jovian system over the course of more than fourteen years. By the 1990s, NASA had adopted a leaner approach to planetary exploration, although it still assembled several significant accomplishments with the *Near Earth Asteroid Rendezvous*, *Mars Pathfinder*, *Lunar Prospector*, *Stardust*, *Mars Global Surveyor*, *Mars Odyssey*, and *Deep Space 1* spacecraft.[26]

During the late twentieth century, NASA also launched three spacecraft that had tremendous impacts on the state of knowledge in astronomy and astrophysics. These "Great Observatories"—the Hubble Space Telescope, *Compton* Gamma Ray Observatory, and *Chandra* X-Ray Observatory—revolutionized these areas of study. The deployment of Hubble in 1990 represented the most significant development in astronomy since the time of Galileo. Orbiting nearly 400 miles above Earth, the space telescope utilized pointing precision, powerful optics, and state-of-the-art instrumentation to provide incredible views of the universe that could not be acquired using ground-based telescopes or other satellites. *Compton*, launched in 1991, had instrumentation that covered an unprecedented six decades of the electromagnetic spectrum, from 30 keV to 30 GeV. During its decade of operations, the spacecraft made fundamental contributions to understanding many classes of galactic and extragalactic objects. *Chandra*, launched in 1999, was the most sophisticated X-ray telescope ever developed. Designed to observe X-rays from high-energy regions of the universe, such as the remnants of exploded stars, *Chandra* produced images of the hot, turbulent regions in space that were twenty-five times sharper than previous X-ray pictures.[27]

In the mid-1960s, President Lyndon Johnson and NASA began searching for potential objectives for the human spaceflight program after the completion of Project Apollo. The space agency desired the adoption of a long-term exploration plan that would include a reusable launch vehicle, orbital space station, lunar outposts, and human missions to Mars. By the time Richard Nixon reached office, however, the space agency had lost a great deal of support at both ends of Pennsylvania Avenue. In March 1970, Nixon decided it was not in the best interest of America to have a high-profile space program and rejected suggestions that the nation proceed with a space station and permanent lunar base in preparation for sending astronauts to Mars. Suddenly, NASA's main concern was with its own survival. As John Logsdon argues, "once NASA's goals in space were rejected, its purpose became maintenance of the institution. A siege mentality developed." It was in this environment that the decision to develop the space shuttle system was made. T. A. Heppenheimer writes, "the Space Shuttle took shape and won support, and criticism, as part of NASA's search for a post-Apollo future. . . . In seeking [that] future, NASA repeatedly had to accept . . . cuts, as its senior officials struggled to win support within the White House."[28]

The space shuttle was originally an afterthought in NASA's grand vision for the future of the American space program. It was intended to provide low-cost flight to orbiting facilities that formed the foundation of crewed lunar and planetary exploration. In the aftermath of Project Apollo, however, NASA was forced to face a harsh reality. Office of Management and Budget (OMB) Director Robert Mayo informed the agency that its FY1970 budget was being cut by $1 billion, which doomed any substantial plans for lunar bases and human exploration of Mars. NASA still intended to go forward with the development of both a space station and a space shuttle, but in the face of congressional skepticism regarding the soundness of this approach, the agency was forced to eliminate the space station from its planning. Bolstered by support from the air force, which believed the shuttle system could launch reconnaissance satellites and other military spacecraft, NASA successfully fought opposition on Capitol Hill and received congressional approval for the program. Despite congressional support, however, the Nixon administration was still not convinced that the shuttle was the proper approach. Pressure from the OMB forced NASA to abandon plans for a shuttle with two fully reusable liquid-fueled stages, compelling the agency to utilize an expendable external tank and refurbishable solid rocket boosters to propel shuttle orbiters into orbit. In early 1972, based on the recommendation of OMB Deputy Director Caspar Weinberger, President Nixon announced that NASA would develop a reusable space shuttle.[29]

After the successful launch of space shuttle *Columbia* in 1981, NASA began to concentrate on obtaining approval to develop a permanently occupied space station. NASA Administrator James Beggs made this goal a central focus of his agenda. In April 1983, on the advice of the National Security Council (NSC), President Reagan authorized a study designed to inform a decision regarding whether to build a space station. NASA became convinced, as the study team conducted its assessment, that it would be impossible to gain support for the station. This was largely due to the reservations of the national security experts, led by Defense Secretary Caspar Weinberger. As a result, NASA decided to attempt an end run and make its case directly to the president. On 1 December 1983, Reagan attended a space station briefing at a meeting of the Cabinet Council on Commerce and Trade. The briefing focused on the contributions the space station would make to bolstering national prestige as well as the commercial applications of the program. The ploy worked. A few days later Reagan approved the Space Station Program, which he announced in his state of the union address the following month.[30]

In the post-Apollo period, the development of the space shuttle and space station represented a large share of NASA's research budget. Throughout this period, however, the agency's budget steadily declined from its mid-1960s peak. By the late 1970s, it had fallen below $15 billion (in FY2018 dollars) annually and barely managed to creep above that level over the subsequent decade. While President George H. W. Bush made a major new commitment to the space program after taking office in 1989, yearly appropriations reached only the $23 billion (in FY2018 dollars) mark during his tenure—still well below the $36 billion (in FY2018 dollars) peak during Project Apollo. During the post-Apollo period the NASA budget dropped permanently below 1 percent of the total federal budget (except for a slight bump in the early 1990s). These continued budget decreases complicated the agency's efforts to develop revolutionary new technologies, although NASA's talented core of scientists and engineers continued to contribute important incremental technical advances in the aviation and space sectors.

## Multiple Streams, Punctuated Equilibrium, and the Life Cycle of Bureaus

This book will utilize two major works within the public policy literature as narrative tools to tell the story of MTPE and its impacts on science and policy. We will leverage the public policy agenda-setting literature, particularly using the multiple streams model developed by John Kingdon and the punctuated equilibrium model advanced by Frank Baumgartner and Bryan Jones. More importantly, we employ the life cycle of bureaus literature, specifically utilizing research conducted by Anthony Downs and James Wilson, to assess the struggles that NASA faced as the space agency evolved its role in earth science research.

In 1972, Michael Cohen, James March, and Johan Olsen introduced the garbage can theory in an article describing what they called "organized anarchies." The authors emphasized the chaotic character of organizations as loose collections of ideas as opposed to rational, coherent structures. They found that each organized anarchy was composed of four separate process streams—problems, solutions, participants, and choice opportunities—and concluded that organizations are "a collection of choices looking for problems, issues and feelings looking for decision situations in which they might be aired, solutions looking for issues to which they might be the answer, and decisionmakers looking for work." Finally, a choice opportunity was "a garbage can into which various kinds of problems and solutions are dumped by participants as they are generated. The mix of garbage in a single can depends on the mix of cans available, on the labels attached to the alternative cans, on what garbage is currently being produced, and on the speed with which garbage is collected and removed from the scene." Therefore, the three found that policy outcomes are the result of the garbage available and the process chosen to sift through that garbage.[31]

In his classic tome *Agendas, Alternatives, and Public Policies*, John Kingdon applied the garbage can model to develop a framework for understanding the policy process within the federal government. He found that there were three major process streams in federal policymaking: problem recognition; the formation and refinement of policy proposals, and politics. Kingdon concluded that these three process streams operate largely independently of one another. Within the first stream, various problems come to capture the attention of people in and around government. Within the second stream, a policy community of specialists concentrates on generating policy alternatives that may offer a solution to a given problem. Within the third stream, phenomena such as changes in administration, shifts in partisan or ideological distributions in Congress, and focusing events impact the selection of different policy alternatives. Kingdon argues that the key to obtaining successful policy outcomes within this organized anarchy is to seize upon policy windows that offer an opportunity for pushing specific proposals onto the policy agenda. Taking advantage of these policy windows requires that a policy entrepreneur expend the political capital necessary to join the three process streams at the appropriate time. We rely heavily on this framework, principally to help shape our narrative approach for this book rather than implicitly testing the model.[32]

The methodological core of the multiple streams model is hundreds of interviews conducted over four years with congressional staffers, upper-level civil servants, political appointees, presidential staffers, lobbyists, journalists, consultants, academics, and researchers. One main objective of these conversations was to determine which players were important in a given policy community. In the early 2000s, Thor Hogan (one of the authors of this book) conducted a survey of the space policy community in an attempt to determine which actors were important within this issue area, and his findings were ultimately published in *Mars Wars: The Rise and Fall of the Space Exploration Initiative.* Because they proved so useful in that manuscript, we decided to make use of the findings to guide the narrative for this book as well.[33]

The president is the single most important actor in setting the agenda within the space policy community. The institutional powers of the office provide its occupant with the ability to heavily influence what policy debates the government will focus on at any given point in time. One advantage enjoyed by the chief magistrate is a large staff working within the Executive Office of the President, the members of which are collectively among the most powerful agenda-setting actors. Staffers working in the OMB, Office of Science and Technology Policy (OSTP), and NSC played outsized roles in setting the path for space policy during the 1980s and 1990s, when MTPE/EOS came to fruition. During the crucial period under President George H. W. Bush, the National Space Council, chaired by Vice President Dan Quayle, joined these other presidential advisers in pushing forward crucial space initiatives. Presidential appointees, such as the NASA administrator, are also among the most frequently discussed actors in the agenda-setting process, at least partially because they have the ability to advise and guide the decisions of the president in an area where chief executives almost never have specific expertise.[34]

It is not possible for a president to dominate the alternative generation process, particularly in an arena where developing policy options requires specific scientific and technical expertise. Luckily for a president, his or her staff is among the most important actors in alternative generation because they do have the requisite knowledge. In addition, political appointees are the most influential actors in finding policy solutions within the space policy community because the NASA administrator can draw upon the expertise of thousands of employees when developing new programs. These career civil servants themselves are key actors in alternative generation because they know more about science and engineering than any of the political players. If political appointees and civil servants work in concert with the president, they provide the executive with a great deal of leverage over the alternative generation process.[35]

Members of Congress have a great deal of influence over agenda setting in the space policy community, although not nearly as much as the executive branch. Within the space sector, congressional staffers are actually more important in agenda setting than the political leaders for whom they work. This significant influence results from the fact that staffers can spend more time than members gaining an understanding of the relevant technical details. Members of Congress are somewhat important actors in alternative generation within the space sector but are not considered to be among the most influential. Representatives and senators are more likely to be involved in shaping specific policy ideas introduced by the executive branch than in developing their own options. Similarly, congressional staffers are not particularly influential in alternative generation. In contrast to other issue areas, where staffers become heavily involved in drafting legislation and negotiating agreements between interested parties, for an initiative such as MTPE/EOS, most legislative aides lack the scientific or technical expertise to develop the very detailed plans needed. This situation has often resulted in mission agencies such as NASA having an inordinate amount of influence over the alternative generation process.[36]

For the most part, nongovernmental actors are not considered to be leading voices in agenda setting within the space policy community. The primary exception is the aerospace industry and interest groups representing those companies, although even they are considered far less important than government actors. Professional and public interest groups, academics and researchers, and the media are all seen as having very limited influence over the direction of the space program. In addition, these nongovernmental actors generally do not have a great deal of impact on the alternative generation process. The aerospace industry is the most influential nongovernmental actor in developing policy options because it has an incredible wealth of knowledge regarding engineering solutions. Interest groups representing the aerospace industry are considered only somewhat influential, while professional and advocacy groups are much less important. Academics and researchers affect alternatives much more than agendas, primarily because they can offer technical expertise that supplements that of civil servants. Although their contributions are commonly ancillary to what happens within government, the evolution of MTPE/EOS proved to be a somewhat radical departure from standard practice. Rather than following the government lead in the development of this program, the earth science community ultimately guided the alternative generation process for it. In the end, that was the only reason that the initiative was successfully adopted.[37]

In *Agendas and Instability in American Politics*, Frank Baumgartner and Bryan Jones introduce a punctuated equilibrium model of policy change in American politics based on the emergence and recession of policy issues on the government agenda. This theory suggests that the policy process has long periods of equilibrium that are periodically disrupted by some instability that results in dramatic policy change. Baumgartner and Jones describe "a political system that displays considerable stability with regard to the manner in which it processes issues, but the stability is punctuated with periods of volatile change." Within this system, they contend that the mass public is limited in its ability to process information and remain focused on any one issue. As a result, policy monopolies are created so that scores of agenda items can be processed simultaneously. Only in times of unique crisis and instability do issues rise to the top of the government agenda to be dealt with independently. At a fundamental level, the punctuated equilibrium model seeks to explain why the policy process is largely incremental and conservative but is also subject to periods of radical change.[38]

For our purposes, Baumgartner and Jones's discussion of the role of policy monopolies proves particularly useful in explaining how MTPE/EOS reached the government agenda. As they suggest, these political subsystems control agendas over indeterminate periods of time that are characterized by stability. During such a period of equilibrium, the policy monopoly attempts to maintain a positive policy image for its approach to dealing with an issue. The policy monopoly also attempts to control the policy venue—the professional, political, legal, bureaucratic, and market-based arenas within which policy discussions take place. The political subsystem often relies on historical precedents to maintain the status quo. Significant changes occur when those without power seek to expand the scope of conflict to change the nature of a policy monopoly's power. Those who are seeking change mobilize criticism, leading to a different focus on the problems, solutions, and policies within a given issue area. A successful challenge takes advantage of different policy images and venues to bring new interest groups, congressional committees, and agencies into the policy debate. These new actors seek to transform the policy debate and establish a new policy monopoly that will enjoy a long period of stability until the next challenge. Baumgartner and Jones refer to this oscillation between stability and change as punctuated equilibrium. Within this construct, agenda setting is the complex process through which policy monopolies maintain control until external actors gain access to that issue area and create a new monopoly. In the final chapter of the book, we utilize the model primarily to examine whether earth science research reached the government agenda, representing the only expansion of NASA's core mission in the agency's history, because external actors successfully challenged existing policy images and venues.

The life cycle of bureaucracies literature proved to be the key tool in establishing a context for NASA's early role in remote sensing and weather research and explaining why the agency's mission was ultimately expanded to include more broad-based earth science research. Anthony Downs wrote one of the first authoritative treatments of administrative agencies in an effort to "develop a useful theory of bureaucratic decision making." Downs suggests that agencies seek to attain their goals rationally; they are motivated primarily by self-interest, and the organization's internal structure is influenced by its social function (and vice versa). Bureaucratic agencies are often created to support a new initiative, which was certainly the case with NASA. Downs argues that all new agencies are staffed by zealots who seek to prove the worth of their bureau and establish its autonomy.[39]

Downs proposes that bureaus have a natural tendency to want to expand. This is particularly true early in a bureau's life as growth provides its leaders with increased power and influence. As this expansion occurs, there is a level of internal competition within the organization. Strongly driven bureaucrats pursue their own interests in an effort to create new divisions within the agency. To control this instinctive inclination, the agency becomes extremely hierarchical. As internal stability is achieved, the organization tends to seek out conflicts with other institutions, which leads to constant jockeying for position and jurisdictional disputes as agencies seek to defend or extend the existing borders of their policy space. In parallel with these ongoing processes, administrative agencies seek to maximize their budget in the same way that private firms seek to maximize profits. Program managers unsurprisingly seek to maximize their divisional budget, leading to general support for a larger overall agency budget. Over the course of an agency's "life cycle," the organization will tend toward more conservative behavior. Downs argues that this tendency does not threaten the bureau, however, because the longer it has been in place and the larger it grows, the harder it is to remove.[40]

More than two decades after Downs's book was published, James Wilson provided an important new perspective on this discussion. While Wilson strongly agrees that agencies seek to ensure autonomy, he suggests that fashioning a core constituency is the key strategy pursued by organizations to ensure long-term stability. He suggests that bureaucrats use six key tactics to craft these dedicated supporters. First, they seek out tasks that are not being performed by other agencies. Second, they fight organizations that seek to perform tasks already performed by their agency. Third, they avoid taking on tasks that differ significantly from those that are at the heart of the organization's mission. Fourth, they are wary of joint or cooperative projects. Fifth, they avoid tasks that will produce divided or hostile constituencies. Finally, they avoid learned vulnerabilities. Wilson contends that these approaches portray administrative agencies that are significantly more complex than the resource-maximizing organizations portrayed by Downs.[41]

Wilson argues further that while it is important to defend existing policy turf, expanding that policy space can be dangerous. Seeking to enlarge the agency's mission introduces a measure of uncertainty because it threatens the agency's autonomy by expanding the relevant policy community. As a result, a larger number of actors will seek to influence the agency as it shapes its policies, making it harder to develop a cohesive sense of mission. Wilson contends that agencies seek to expand their turf only when they feel reasonably assured that their autonomy will remain intact. This book seeks to evaluate whether this argument holds true when investigating NASA's efforts to expand its original mission to include earth science research.

This book is the first to describe and analyze the evolution of the MPTE/EOS initiative from its formative years in the 1980s to its political and technical struggles in the 1990s, primarily using a policy lens to develop the narrative. In this way, it fundamentally differs from much of the existing literature that focuses more on scientific or technical developments. There are a limited number of scholarly treatments of the evolution and successes of NASA's earth science programs, in contrast to many human spaceflight programs. This is unfortunate given the importance of this research within the contemporary social and political environment. However, a number of excellent books, reports, and dissertations tell parts of this story and were crucial to our research. Before discussing a broader array of sources, our hope is that this book will be seen as a meaningful complement to Erik Conway's superlative *Atmospheric Science at NASA: A History.*[42] This award-winning book took a critical first step in telling this overarching story, although we believe this manuscript is a compelling extension. While Conway focuses much of his attention on atmospheric science and the development of climate change research, we concentrate far more deeply on the executive branch politics and bureaucratic infighting over administrative control, satellite design, and budgeting that led to the ultimate adoption and revision of MTPE/EOS. Although there are overlaps between the two books—in particular, both cover the development of the overarching system architectures for MTPE/EOS—they essentially tell the same story using different scholarly lenses. We believe there is great value in this duality of approaches, particularly as there is a need for more books that promote the importance of NASA's earth science efforts.

Beyond Conway's book, there are a number of scholarly sources that, together, provide a detailed history of NASA's weather and remote sensing satellite programs during the 1960s and 1970s. A good starting point is a doctoral dissertation written by Richard LeRoy Chapman titled *A Case Study of the U.S. Weather Satellite Program: The Interaction of Science and Politics*,[43] which is an excellent reference for gaining an understanding of the space agency's early weather satellite programs as well as insights into the early links between science and politics. A helpful accompaniment to this book is Janice Hill's *Weather from Above: America's Meteorological Satellites*.[44] Without a doubt the best book in this category is Pamela Mack's *Viewing the Earth: The Social Construction of the Landsat Satellite Program*,[45] which is the definitive history of the early days of the Landsat program. The book traces the development of Landsat from its origins through the launch and use of the first few satellites and provides valuable insights into the roles played by various governmental and nongovernmental actors in shaping new technology.

Based on this foundation, we utilized a number of other works as secondary sources when discussing the evolution of earth science research at NASA during the 1980s and 1990s. While these works vary in comprehensiveness, overall they provide a useful patchwork history of this period. Perhaps the best introductory sources in this arena are John McElroy and Ray Williamson's essay "The Evolution of Earth Science Research from Space: NASA's Earth Observing System" in *Exploring the Unknown: Space and Earth Sciences*[46] and a dissertation by Edward Goldstein at The George Washington University.[47] In many respects, these served as essential primers for understanding the crucial events in the development of an earth science mission at NASA. Additionally, the primary sources accompanying both were invaluable resources. Likewise, there is an extremely useful National Research Council (NRC) report called *Earth Observations from Space: History, Promise, and Reality* that provides an overview account of a good portion of the history that we cover.[48] A number of other important reports also helped highlight the changing nature of the debate in this policy area over time:

- *Global Change: Impacts on Habitability* was the first report stating that global environmental studies should focus specifically on the effects of global changes on habitability and is credited with identifying NASA as the critical player in a global study of the environment;[49]

- *A Strategy for Earth Science from Space, Part I: Solid Earth and Oceans* was the first of two NRC Space Studies Board reports to provide a comprehensive review of the important scientific questions to be addressed by satellite studies of the earth;[50]
- *Toward an International Geosphere-Biosphere Program: A Study of Global Change* was the final report of a panel chaired by Herbert Friedman that was the first to discuss strategies for adopting an in-depth international earth science research program based on understanding the long-term changes of the Earth system;[51]
- *Leadership and America's Future in Space*, a report to the NASA administrator by Dr. Sally Ride, known as the Ride Report, was undertaken in response to the National Commission on Space and listed MTPE/EOS as one of the key future programs for the space agency;[52]
- *Earth System Science: A Closer View: A Program for Global Change* was the second of two highly influential "Bretherton Reports" that set the stage for NASA's involvement in this research area;[53]
- *Our Changing Planet: A U.S. Strategy for Global Change Research* was the first publication by the George H. W. Bush administration on a coordinated federal program in the global earth sciences and was produced annually for several years;[54]
- *Report of the Advisory Committee on the U.S. Space Program*, known as the Augustine Report, was a highly respected review of NASA that strongly reiterated the Ride Report's conclusions that a science program, particularly MTPE/EOS, should be a major thrust for the space agency;[55] and
- *Report of the Earth Observing System (EOS) Engineering Review Committee*, the report of the Frieman Committee to "restructure" EOS, was the first report to kick off several years of EOS redesigns.[56]

In addition to these influential reports, several additional scholarly works deserve mention because they were important sources for the book. One of the key books that provided context for the political environment facing MPTE/EOS during its formative years was Howard McCurdy's *The Space Station Decision: Incremental Politics and Technological Choice*,[57] which was important for understanding the early link between the Space Station Program and earth science research. Finally, an excellent article by Henry

Lambright, "The Ups and Downs of Mission to Planet Earth," tracked the tumultuous development of MPTE from the 1980s through the mid-1990s.[58]

Beyond the secondary literature, we conducted extensive primary research. We took particular advantage of the archives at the NASA History Office in Washington, D.C., and to a lesser degree drew on materials from the George H. W. Bush Presidential Library in College Station, Texas. In addition, we conducted numerous interviews with the key actors involved in the struggles to gain approval for the MTPE/EOS initiative during the 1990s. We also relied heavily on scores of additional interviews conducted by a number of other authors, including an article by Edward Edelson,[59] two case studies by Harvard University's David Kennedy,[60] and, a dissertation by Roger Pielke at the University of Colorado.[61] These sources provided invaluable additional insights into the relationship between the U.S. Global Change Research Program, Congress, and NASA.

Building from these sources, the aim of this book is to answer several fundamental questions regarding NASA's adoption of MTPE/EOS. First, in the early 1980s, why did NASA choose to give only tepid support to the inclusion of an aggressive earth science research program on the national agenda? Second, why did the space agency, which had not been keen on battling for a new initiative in the first place, spend the remainder of the decade pursuing it? Third, in 1989, how did the initiative reach the national agenda and gain formal adoption? Finally, during the bureaucratic battles that ensued in the 1990s, how did NASA maintain support for the program as political and budgetary priorities changed? In answering this final query, we will consider how the space agency resolved numerous critical challenges. These disputes included whether MTPE/EOS would be based upon a few large platforms or a larger number of smaller satellites, whether the satellites would be expendable or serviceable in orbit (by shuttle crews), and whether the scientific parameters and satellite architectures would be designed by NASA or shaped by the scientific community. In examining the outcome of these various clashes, we will seek to provide a better understanding of how science, technology, and policy converged in one of the most significant Big Science programs in human history. This is particularly germane in an era when rapid scientific and technical advancements are increasingly shaping human society.

In the end, we find that NASA succeeded with MTPE/EOS primarily because it had a pragmatic and assertive client constituency that helped it navigate the fraught political landscape of the 1990s. While the literature on aging bureaucratic agencies suggests that they are generally incapable of winning support for new missions, NASA benefited greatly from its strong relationship with a science community that was willing to essentially force the agency to develop MTPE/EOS in a way that comported with the contemporary political reality. The necessity to play client politics was, therefore, not a burden but a tremendous windfall. NASA pursued this initiative largely because the science community promoted the idea, and it was successful in gaining adoption of the program because of the tough love that forced it to reenvision MTPE/EOS in the years after the program was initially approved.

The initial chapters of this book will examine the events leading up to the adoption of MTPE. The central focus of this story is a detailed account of the agenda-setting process that placed the program on the government agenda. Several of the following chapters will explore the political battles that were fought over the subsequent decade to ensure that the initiative would actually be implemented. Finally, the concluding chapter will consider how the models introduced above can provide us with a better understanding of the events portrayed in earlier chapters. This analysis will begin with a discussion of how the multiple streams and punctuated equilibrium models can help explain how and why MTPE/EOS reached the national agenda and was ultimately adopted, followed by a lengthier consideration of how the life cycle of bureaus literature can provide us with a deeper understanding of how NASA's strong external client constituency helped it add a significant earth science mission to its existing portfolio of programs. The final section of the chapter will consider the central lessons that emerge from our research into the history of the MTPE/EOS program.

## CHAPTER ONE

# A Global View of the Earth

Remote sensing, the scanning of the earth by high-flying aircraft or satellites, became technologically possible with the invention of the camera nearly two centuries ago. During the 1800s it was possible, although incredibly rare, for someone to ascend via a balloon and snap photographs to study the landscape below. In this way, topographic maps could be produced. After Orville and Wilbur Wright built the world's first successful powered airplane, remote sensing became one of the earliest practical applications for the new technology. In the course of World War I, military strategists began to appreciate the benefits of aerial photographs. Cameras were designed specifically for this purpose, although they remained fairly basic. After the war, the U.S. Army Air Service began to conduct extensive aerial mapping surveys and helped establish a small civilian aerial photography industry. Photogrammetry, the study of photographs to make accurate measurements of the physical world, emerged during the interwar period as new instruments were invented. In the 1930s, the applications began to extend beyond the military into civilian life, particularly environmental management and economic planning. During the Great Depression, aerial photography was used as part of the federal government's programs for agricultural recovery and erosion control, and the Tennessee Valley Authority used it to help plan flood control efforts. In 1934, this collection of American government programs eventually gave rise to a new professional organization called the American Society of Photogrammetry, which published an important academic journal named *Photogrammetric Engineering and Remote Sensing.*[1]

During World War II, photographic technologists moved beyond the visible spectrum into the electromagnetic spectrum and began developing suitable applications. Aerial remote sensing played a far larger role in this conflict, with many more personnel involved in the process. Military leaders were able to go beyond using these capabilities solely for tactical decisions and could use the information gathered to make fundamental strategic determinations. This was particularly important in assessing the enemy's infrastructural and industrial strengths and weaknesses. As early as 1946, the U.S. Air Force (USAF) began working with the RAND Corporation to consider how an orbiting spacecraft might function and to determine the potential benefits offered by such a platform. The initial report on a "world-circling spaceship" identified both military and scientific uses. It concluded that the evolution of satellites was too speculative to calculate the possible benefits—although RAND did suggest that the benefits would likely be grander than those foreseen by early aviation pioneers. While budget pressures and the dominance of nuclear weapons research delayed plans to build a world-circling spaceship, RAND was directed to continue studying the potential utility of Earth-orbiting satellites.[2]

In 1950 the think tank submitted another report to the DOD that Walter McDougall argues was the "birth certificate of American space policy." The study was a treatise on the potential national and international reaction, both political and psychological, to the launch of the first satellite. Paul Kecskemeti, who authored the report, argued that it would be in America's interest to launch such a satellite. Noting that satellite reconnaissance was likely the best application of the technology, he reviewed the legal complications that could derail such a project and suggested strategies to overcome those obstacles. The report was exceptionally well received, but the race for ICBMs and the Korean War continued to keep satellite development on the back burner. Nonetheless, Kecskemeti had proven to many the worth of such a program, and RAND was directed to develop more specific concepts and detailed satellite designs.[3]

After the launch of *Explorer 1* during the IGY proved that the nation could successfully orbit satellites, military planners accelerated their efforts to develop reconnaissance satellites that would use both photographic technology and nonphotographic sensors (e.g., radar to improve missile target selection, infrared to provide heat information). To interpret the resulting data, the DOD turned to the same geographers and geologists who had previously proven to be expert in evaluating aerial photography and photogrammetry. Although nearly all defense projects were classified, the DOD worked closely with the larger scientific community to develop remote sensing.[4]

## NASA's Weather Satellite Programs

As Project WS-117L progressed during the late 1950s, the USAF began the development of the Satellite Military Observation System (SAMOS) to test visual surveillance using scanned photographs radioed to the ground. At the same time, ABMA was considering a Radio Corporation of America (RCA) proposal to build a similar satellite using a television camera, which ultimately led to the adoption of the Janus program. In early 1958, however, the DOD decided that satellite reconnaissance would be the operational responsibility of the USAF, which meant that all ongoing projects were either canceled or transferred to this branch. Still trying to gain a foothold in space, the ABMA sponsored Janus II. This project used the same technology as Janus but focused on meteorological applications.[5]

Indicative of the administrative juggling that characterized this early period of space history, the Janus II project was almost immediately transferred to the newly created Advanced Research Projects Agency (ARPA). The new agency quickly changed the name of the program from Janus II to Television Infrared Observing Satellite (TIROS). The TIROS satellite was designed to be a polar-orbiting satellite operating in a low-Earth, sun-synchronous orbit that would pass over the equator at the same time every day. In April 1959, the project was transferred yet again, this time to the newly created NASA, which had been tasked by President Eisenhower with taking over DOD space programs that were not clearly military in nature. Thus, the new civilian agency was given responsibility for TIROS and directed to develop the satellite in cooperation with the Weather Bureau.[6]

By the time NASA took control of TIROS, the initial spacecraft design was already complete, and prototypes were under construction. On 1 April 1960, the first TIROS satellite was launched and began returning photographs to the earth. The satellite had limited capabilities, as its cameras could face the earth for only half of its orbit, and it provided no coverage of the polar region. Nevertheless, it was remarkable for its time and produced amazing results that were shared with Eisenhower and Congress. TIROS was immediately heralded as a success, even though it suffered an electrical system failure after only two months. In November 1960, TIROS 2 was successfully propelled into orbit, and over the subsequent half decade, a total of ten TIROS platforms were launched. These missions were central to demonstrating the usefulness of meteorological satellites. With this experimental program well established, NASA and the Weather Bureau began investigating operational weather satellites.[7]

In October 1960, NASA, the DOD, and the Weather Bureau partnered to form an interagency Panel on Operational Meteorological Satellites. It did not take long for bureaucratic conflict to arise around issues such as satellite design and operational responsibility. The Weather Bureau lobbied for complete authority over the operational system, including final decisions on the design of new satellites. NASA was not willing to part with this much control. The panel ultimately reached a compromise that actually gave NASA most of what it wanted. The Weather Bureau would define the requirements for weather satellite design and the distribution of the resultant photographs, and NASA would maintain overall control of spacecraft design, development, procurement, launch, communications, and day-to-day operations. In effect, NASA would have near-total authority over operations. The panel recommended the development of a program called the National Operational Meteorological Satellite System (NOMSS), which would be based on NASA's design for a second-generation weather satellite named Nimbus. In January 1962, with support from both President Kennedy and Congress, the three agencies signed an agreement for NOMSS, with the first Nimbus launch planned for later that year.[8]

The Weather Bureau was never comfortable with Nimbus as the TIROS replacement, primarily because the new satellite design was significantly more complex and difficult to develop. This concern was validated only six months after NOMSS was created, when NASA announced that the launch of the first Nimbus satellite would slip until at least mid-1963. The biggest fear was that this delay would cause a gap in coverage between the two satellite programs. To compensate, NASA extended the TIROS program, agreeing to launch an additional eight spacecraft, but Nimbus delays still threatened to leave a coverage gap. This led to ongoing bureaucratic infighting between the Weather Bureau and NASA. The two agencies could not agree on basic technical details (e.g., the proposed orbital altitude) or management (e.g., which agency would receive the data transmitted by the satellites). In September 1963, this battle reached the boiling point when the Weather Bureau announced its intention to withdraw from all interagency agreements with NASA and pursue its own operational satellite program (using technology based on TIROS). In addition, the Weather Bureau recommended to the White House budget office that Nimbus be canceled.[9]

The tactic worked, forcing NASA back to the negotiating table to reach a more balanced compromise. NASA would develop (using funding from both agencies) an operational satellite based on TIROS technology using requirements established by the Weather Bureau. The Weather Bureau would make final design decisions and establish cost and schedule parameters for satellite development. NASA would provide launch services but would immediately cede control of each satellite to the Weather Bureau once the platform successfully reached orbit. From that point on, the Weather Bureau controlled all program operations. In January 1964, the two agencies formalized this new arrangement in the TIROS Operational System (TOS). It was agreed that the DOD would continue its own meteorological satellite program called the Defense Meteorological Satellite Program (DMSP). Furthermore, the Nimbus program would continue under NASA control as an experimental program intended to advance new instrument technologies relevant to weather and Earth observations. If these new technologies were found to be useful, the Weather Bureau would decide whether to fund their inclusion in new operational satellites. The TOS program was a welcome change in the relationship between NASA and the Weather Bureau (which was transferred to the Environmental Science Services Administration in 1965).[10]

While the TIROS and Nimbus satellites were polar-orbiting spacecraft, the NOMSS plan also called for an operational geosynchronous satellite. During the mid-1960s, NASA initiated the Applications Technology Satellite (ATS) program to meet this goal and to demonstrate the usefulness of satellites for navigation, communications, and meteorology. In December 1966, the *ATS 1* spacecraft was launched into orbit carrying communications and meteorological experiments. The meteorological sensor, the Spin-Scan Cloud Camera (SSCC), was designed to take a series of photographs of the earth as the satellite spun to keep itself stable and was able to capture an image of most of the globe from north to south.[11]

In 1970, after nearly five years of bureaucratic stability, the Department of Commerce was reorganized once again by an executive order signed by President Nixon. The newly created National Oceanographic and Atmospheric Administration (NOAA) absorbed the Environmental Science Services Administration and a number of other disparate organizations. Three years later, NASA and NOAA concluded a new cooperative agreement called the Operational Satellite Improvement Program (OSIP). Under OSIP, NASA would fund "first unit" builds of weather satellites and instruments and transition the proven capabilities to NOAA for operational use. The agreement reflected the continuing basic understanding that the two agencies would maintain their respective roles in development and operations. During subsequent years, NASA focused its research funding on the development of prototype sensors that were tested on high-altitude aircraft. Successful sensors were then transitioned to research spacecraft for evaluation. Ultimately, successful instruments were made available to NOAA for operational spacecraft.[12]

In 1964, NASA had launched the first Nimbus (which means "rain cloud" in Latin) spacecraft. It was intended to be a research tool for improving future operational satellites. At the time, there were many ideas about how best to improve weather prediction with space technologies. Photographs of cloud formations, provided by platforms such as TOS, were only somewhat useful. Although they improved hurricane forecasts, they did not help with predicting weather conditions outside the coastal regions. Additionally, these satellites did little to advance the Weather Bureau's efforts to develop weather prediction models that would provide better forecasts using accumulated wind and temperature data. Radio astronomers had previously devised techniques to calculate atmospheric and surface temperatures using the electromagnetic emissions of other planets. Some contemporary meteorologists thought this form of remote sensing might be useful for their own research. Others within the field believed that satellite-tracked balloons could also provide the global-scale data sets needed for weather prediction researchers. One of the most important objectives of the Nimbus program was to investigate as many of these new technologies as possible to provide new sensors for researchers in meteorology and other earth science disciplines.[13]

The *Nimbus 1* spacecraft was the first weather satellite to be stabilized without spinning. This allowed it to continuously correct its orientation and always point its instruments toward the earth. It survived for only one month, however, as it tumbled out of its orbit after its solar panels accidentally locked. In May 1966, *Nimbus 2* was launched carrying a suite of three innovative instruments—the Advanced Vidicon Camera System (AVCS), the High-Resolution Infrared Radiometer (HRIR), and the

Medium-Resolution Infrared Radiometer (MRIR). The HRIR produced images that made warm surface areas look dark and cold areas look light, helping highlight the coldest elements of the atmosphere—high-altitude clouds. The sensor was also the first to provide readings of sea-surface and sea-ice temperatures, was able to gather images of cloud tops at night, and revealed the outlines of the major ocean currents. The MRIR was the first microwave instrument placed on a satellite. It collected data on Earth's incoming and outgoing heat radiation, thus establishing measurements of the planet's radiation budget and providing valuable baseline evidence for early climate researchers. The instrument's data demonstrated that water vapor distribution in the atmosphere could be mapped from space and showed that, as most atmospheric scientists had expected, carbon dioxide was uniformly distributed in the troposphere. *Nimbus 2* succeeded in demonstrating several important capabilities, and its instruments eventually found their way onto operational NOAA satellites.[14]

Three years later, *Nimbus 3* was launched with a larger number of advanced instruments aboard. These included the Satellite Infrared Spectrometer (SIRS) and the Infrared Interferometer Spectrometer (IRIS), which were the first sounding instruments in space. They were able to detect water vapor at different altitudes and the ozone content in the upper atmosphere. The spacecraft was an important precursor to the Improved TIROS Operational Satellite (ITOS), Geostationary Operational Environmental Satellite (GOES), and NOAA satellites, and several of its sensors were ultimately included on those NOAA platforms. Over the next decade, three more Nimbus spacecraft continued to improve upon the measurement capabilities of operational satellites.[15]

## NASA, the USGS, and the Battle for Landsat Approval

When NASA was created, many within the scientific community hoped that satellite remote sensing would be a major focus for the new civilian program. The space agency, however, focused its earth observation programs on meteorological research. At the time, President Eisenhower expressed a clear preference that satellite reconnaissance be the exclusive purview of the DOD. Nevertheless, within a few years, military research aimed at advancing aerial photography toward remote sensing had given rise to a new community of scientists and industry professionals who were interested in space-based observations to monitor natural resources and to pursue other clearly civilian scientific objectives. In February 1962, the

Environmental Research Institute at the University of Michigan organized the First Symposium on Remote Sensing of the Environment. The inspiration for the conference was work conducted for the DOD by the institute's Infrared Laboratory to develop battlefield surveillance techniques. The researchers had realized that many of the techniques advanced under this contract also had nonmilitary applications. Sponsored by the Office of Naval Research, this initial meeting featured 15 presenters and 70 participants (a second symposium held eight months later drew 162 participants to hear 35 presenters).[16]

These meetings led to the first civilian remote sensing experiments as part of the Mercury human spaceflight program. In October 1963, astronaut Walter Schirra carried a hand-held 70-mm camera aboard his *Mercury 8* spacecraft to take photographs of the earth's surface. This project was repeated seven months later by astronaut Gordon Cooper aboard his *Mercury 9* spacecraft. These two missions produced over fifty terrain photographs that were useful to researchers investigating earth resource–surveying techniques. The success of this effort led to the adoption of the Synoptic Terrain Photography Experiment, which was carried on Project Gemini flights starting in 1965. The ten Gemini missions produced over 1,100 color images that were useful for geology, geography, and oceanography research (these photographs only increased the desires of remote sensing scientists for more advanced technologies). As they were produced, the images were published in several magazines, including *National Geographic.* They generated a great deal of international interest in the potential value of orbital imagery of the earth's surface beyond meteorology.[17]

While the photographs taken by NASA astronauts received a great deal of public attention, they were only part of a larger earth resources research program being conducted by the Office of Manned Space Science (OMSS). Between 1963 and 1966, this office sponsored extensive remote sensing research projects in two areas. First, academic researchers were contracted to conduct investigations regarding the applicability of data collected in the nonvisible spectrum to agriculture, geology, hydrology, and other scientific disciplines. Some of the funding for these research projects was transferred to other federal agencies (e.g., the U.S. Department of Agriculture [USDA] and Department of the Interior [DOI]) so that the federal government could establish a broad user base for remote sensing data. Second, academic and industry researchers were contracted to develop more advanced sensor technologies—specifically, sensors in the nonvisible light spectrum. At the same time, the OMSS was conducting an aircraft research program that flew the first versions of these sensors at high altitudes. In early 1966, NASA conducted studies of three classes of potential satellites for earth observation—small, medium, and large. The OMSS favored large satellites that would carry people and multiple instruments as part of the Apollo Applications Program. It also favored the construction of an initial satellite that would test and validate a suite of eighteen sensors: four camera systems, three radar systems, six infrared and microwave systems, and five other advanced sensors. It was hoped that the success of these tests would support the development of a more permanent crewed facility.[18]

That summer, RCA officials approached scientists at NASA's Goddard Space Flight Center with an idea for a small satellite mounted with a camera dedicated to terrain photography for geological research. The RCA proposal did not fit with NASA's plans, so the company was directed to officials at the U.S. Geological Survey (USGS). Researchers at the USGS were convinced that remote sensing data from space would be beneficial to the academic and industrial constituencies that they supported. They were also impatient with the slow progress of NASA's aircraft testing program and the failure to quickly develop an operational satellite. In September, the DOI—the department that housed the USGS—announced its intention to move forward with its own Earth Resources Observation Satellite (EROS) program, which would focus on mapping and geology research. The USGS planned to launch a satellite within three years, which would make the DOI the lead agency for land remote sensing research. Although the White House ultimately rejected the DOI proposal, it did support a joint satellite program with NASA.[19]

In October, the USGS communicated to NASA its requirements for a remote sensing spacecraft. It proposed that the primary instrument would be a television-like camera similar to those flown on TIROS satellites. The primary interest was in fine-resolution images that could easily be compared to maps and aerial photographs for geological and natural resource studies. The expectation was that such a relatively simple technology could be launched within three years. The USGS wanted "an operational satellite as soon as possible so that they could justify their interests by using the data in the routine execution of the missions they had been given by Congress." It took over six months for NASA to officially respond. The space agency did not support such an operational system, instead favoring an experimental satellite that would test important sensors, data storage methods, and data transmission technologies before developing a complete system. During the ensuing months, NASA conducted a concept study for an earth resources satellite based on this plan.[20]

Despite NASA's initial reticence, the study resulted in the recommendation of a satellite very similar to that proposed by the USGS the previous year—a small satellite based on the Nimbus design with a limited number of sensors and developed quickly. The primary reason that the space agency reached this tardy decision was the conclusion by the study team that such a satellite was fiscally and technically feasible. Therefore, NASA and the USGS agreed upon a working arrangement very similar to that of the space agency and the Weather Bureau. NASA would lead the satellite development phase, while the USGS would operate the satellite once it was launched into orbit. In fall 1967, the two agencies included EROS (eventually renamed Landsat) in their FY1969 budgetary requests. The USGS also requested funding for a proposed Environmental Resources Observation System Data Center to complement future platforms.[21]

The two agencies faced an uphill battle, as the White House Bureau of the Budget (BOB) was not inclined to support applications for programs before they had been proven practical. While it may have been possible for NASA to gain approval to fly experimental Landsat sensors on Nimbus satellites, the agencies were asking the BOB to commit to a new operational program. This met with resistance, particularly because NASA planned to continue directing a great deal of Landsat funding to air-based sensors that would seek to mature these technologies. For many at the BOB, this left the potential benefit of space-based sensors in doubt. If the sensors could be flown on aircraft, why fly them in space? As a result, NASA's FY1969 budget request for the project was cut from $25 million to $8 million. The USGS fared even worse—its $3 million request was slashed to only $200,000. NASA Administrator Jim Webb appealed the decision directly to President Johnson and was able to obtain an additional $4 million (the USGS did not have similar clout at the White House, so its funding cut remained intact). Although NASA was able to secure $12 million, the BOB directed the space agency to work with the USGS to conduct a cost-benefit analysis of the Landsat satellite. The study had to be conducted within a year and prove that the program's operational benefits were sufficient to justify a large expenditure of taxpayer money. The following year, NASA and the USGS submitted the requested report to the BOB. Conducted by the Westinghouse Corporation, the study was extremely well received by the Johnson White House. In a memorandum reviewing the study, a BOB staffer noted that "we believe the case for potential economic benefits has been established." The BOB allotted $25.1 million for Landsat, enough to support a development contract for the first satellite, which was targeted for launch in 1972.[22]

Unfortunately for NASA and the USGS, the entire cycle began again after Richard Nixon took office. The Nixon administration reopened many of the previous year's debates over the costs and benefits of Landsat and the potential for aircraft programs to do a better job. In early 1969, the BOB once again imposed severe reductions on the Landsat budget and demanded that the space agency prove the economic benefits of the program. This time, NASA Administrator Thomas Paine (who had replaced Webb) saved Landsat by appealing directly to Nixon. In October, NASA was able to begin a contract competition between General Electric and TRW. The fight between NASA and the OMB—the successor to the BOB—continued when budget officials challenged several of the technical elements and anticipated benefits of the Landsat program. The space agency was forced to walk a thin line. While arguing that many of the tangible benefits from these new technologies were simply unknown, it was still making the case that there was a clear cost-benefit justification for the program. NASA was able to successfully navigate these recurring budgetary challenges and keep the program on schedule. In July 1972, *Landsat 1* was successfully launched into orbit. During that same year, the agency secured funding for a second spacecraft, and *Landsat 2* was launched three years later.[23]

## The Evolution of a Global Perspective in the 1970s

Shortly after *Landsat 1* was launched, several of its sensors failed. Luckily, a new instrument that had been designed specifically for the platform exceeded expectations. For decades, the NASA aircraft program's ongoing study of new sensor technologies had led geologists, geographers, and foresters to create a spectral reflectance library of the earth's features (grass, mountains, urban areas, plants, and many varieties of trees). By the time *Landsat 1* was launched, the scientific community had an established database of how the earth's features would look through the lens of a multispectral instrument. Using this information, *Landsat 1*'s multispectral scanner (MSS) provided colorful images of farming areas, forests, cities, mountain ranges, and even highways. Though the relatively low-resolution (80 meters) multispectral (four color bands) data collected did not reveal much detail, the images were considered highly useful for documenting large-scale changes in vegetation such as deforestation. Two years after *Landsat 1* was launched, NASA won the Collier Trophy, awarded annually for the greatest achievement in aeronautics and astronautics, for "proving the value of U.S. space technology in the management of the Earth's resources and environment for the benefit of all mankind."[24]

The data acquired by the *Landsat 1* and *Landsat 2* spacecraft provided a new global view of the earth's land cover. As a result, the data found almost immediate use in scientific research, including a variety of fundamental geological investigations (e.g., mineral and rock identification), monitoring and cataloging renewable energy resources, tracking the movement of water on land (e.g., rivers, lakes, snow cover, and ice masses) to learn more about the hydrological cycle, and measuring the balance between urban land use and natural land cover. In addition to providing global coverage, the Landsat satellites also supplied another feature that aircraft measurements could not—repeatability. The satellites operated in a near-polar sun-synchronous orbit, passing over the same exact spot every eighteen days at the same time of day. As a result, researchers could add a new time dimension to their environmental measurements. During the early years of Landsat operations, this presented NASA scientists with the capability to begin tracking environmental health and environmental change. For example, researchers were able to measure how large vegetative areas change with the seasons and how complex interactions between oceans and coastlines function over time.[25]

During the late 1970s, NASA began efforts to expand the Landsat program. The agency built receiving stations in ten countries around the world, providing backups for the system's onboard data recorders and enabling Cold War allies to play a role in the American remote sensing program. In addition, NASA and the USGS developed working relationships with other government agencies to generate an improved understanding of the value of Landsat data for agriculture, forestry, land management, and hydrology. Two such programs were the Large Area Crop Inventory Experiment (LACIE; a cooperative program with the USDA and NOAA) and the Agriculture and Resource Inventory Surveys through Aerospace Remote Sensing (AgRISTARS; a cooperative program with the USDA). NASA, the USGS, and the USDA also labored to prove that Landsat data could be used by federal and state agencies to comply with new laws requiring environmental monitoring. Finally, NASA sponsored several technology transfer and utilization programs involving universities, industry, and state and local governments. All of these efforts, combined with the scientific benefits provided to that point, were sufficient to gain approval for a third satellite—*Landsat 3* (launched in 1978).[26]

In November 1967, NASA launched *ATS 3* (*ATS 2* had experienced a launch failure several months earlier). This experimental weather and communications spacecraft carried a color version of the SSCC, which produced the first full-disk color photograph of the earth from space one year before the more famous image snapped by the *Apollo 7* astronauts. The success of *ATS 3* was enough to gain support for continued improvements in this satellite family. In 1970, NASA awarded a contract for the development of the *ATS 6* satellite (after two more launch failures) and issued a new contract to build an upgrade to the ATS design. Calling the improved platform the Synchronous Meteorological Satellite (SMS), NASA agreed to build two such satellites. Despite all of their successes, one of the problems with the TIROS and ATS satellites launched during the 1960s was that they provided only two-dimensional information about the atmosphere. For example, they produced photographs that did not supply any data about the depth of the atmosphere and could not differentiate between phenomena occurring at various altitudes. For earth scientists, the Nimbus and SMS platforms launched in the subsequent decade were changing the whole nature of human knowledge about weather and climate by contributing the information they needed about the vertical dimension of the atmosphere. This aided in the development of advanced mathematical models for weather forecasting based on, and validated by, the observed data. A unique feature of these geostationary satellites was their ability to examine a cross-section of the atmosphere at the edge of the images. In these areas, scientists could measure the effects of pollutants, as they changed how light passed through the atmosphere. This data was critical, as scientists were slowly becoming more aware of the greenhouse effect and global climate change.[27]

*Nimbus 4* carried an updated IRIS, which was the first instrument to demonstrate that an interferometer could measure temperature profiles at different atmospheric elevations. This data, in turn, could be employed to assess the radiance of carbon dioxide in the atmosphere. Thus, the *Nimbus 4* IRIS made possible the first observational determination of a "greenhouse effect" in the earth's atmosphere, building on terrestrial observations by Charles Keeling of carbon dioxide concentrations in the atmosphere. A growing body of literature was suggesting that the planet was warmer than it would have been without carbon dioxide in the atmosphere. This research was based on theoretical calculations and laboratory measurements of the radiative characteristics of various gases found in the atmosphere. However, these radiative properties had never been measured in the actual atmosphere. The IRIS instrument was able to make the necessary measurements of these gases, including carbon dioxide, and thus provided key data for the theoretical models being developed at the time. Both IRIS and SIRS utilized the infrared spectrum, but neither instrument could produce temperature profiles below clouds because cloud cover almost completely absorbs infrared radiation. The result was errors in numerical forecasts using data from the instruments. This shortcoming led researchers to begin investigating new instruments that could penetrate most types of clouds, which guided them toward microwave radiation. Atmospheric oxygen radiates in the microwave region, and meteorologists believed this would permit the measurement of temperatures in the same way that carbon dioxide infrared radiances were assessed. The Nimbus E Microwave Spectrometer (NEMS) instrument on *Nimbus 5* was a microwave-sounding instrument able to measure vertical atmospheric conditions regardless of cloud cover. *Nimbus 6* carried an even better instrument called the Scanning Microwave Spectrometer (SCAMS). The measurements from these instruments were highly useful to weather and climate modelers. As a result, they were quickly added to forthcoming TIROS-N weather satellites. Nimbus did more than just demonstrate technologies that were useful for operational weather forecasting. Instruments aboard these spacecraft also provided scientists with previously unknowable information that enabled a new wave of study of global processes in the earth's environment: the movement of water vapor in the jet stream, the movement of dust and volcanic ash through the atmosphere, the fluctuation of ozone levels over highly populated urban areas, the changes in sea ice and sea temperatures at the poles, wind velocity at different altitudes in the atmosphere, changes in tropical wind circulation, the movement of global ocean currents, and fluctuations in the earth's total radiation budget.[28]

In 1972, NASA's Langley Research Center hosted a conference on the potential uses of remote sensing technologies for carrying out pollution and climate-related research. The study that was produced by the conference participants, *Remote Measurement of Pollution*, detailed the techniques that might be used to measure a wide variety of atmospheric substances, including nitrogen oxide, carbon dioxide, sulfur dioxide, ozone, and fluorocarbons. NASA was establishing itself within the climate and environmental science community as the agent in the government with the necessary technical know-how to address their problems. This conference began the space agency's shift toward environmental research. Utilizing an opportunity to establish itself on thus far unclaimed turf, NASA was enabled by the meeting to define the *Nimbus 7* satellite as a *pollution patrol* satellite. While the satellite would retain its meteorological functions, NASA designed it to carry a number of instruments aimed at detecting various pollutants, and its instruments were the first intended to examine questions related to stratospheric ozone depletion.[29]

In October 1978, NASA launched *Nimbus 7*, which was the last spacecraft in this series. It was highly advanced for its time, with instruments to measure and observe radiation reflectance off clouds and absorption in the oceans; the presence and movement of aerosols in the atmosphere; a larger share of the electromagnetic spectrum; gases and particulates in the atmosphere to determine the detailed behavior of atmospheric pollutants; ocean color, temperature, and ice conditions (particularly in coastal zones); and radiation fluxes outside the atmosphere and of atmospheric constituents. What was most remarkable about *Nimbus 7* was that some of its instruments, such as the Stratospheric and Mesospheric Sounder (SAMS) and the Limb Infrared Monitor of the Stratosphere (LIMS), were paired in such a way as to provide coincident measurements of the same point in the atmosphere. This "provided synergy among multi-instrument payloads . . . and for the first time, a variety of powerful, complementary techniques were used to measure geophysical parameters and understand Earth system processes." The Nimbus satellites were also the first to widely distribute data as soon as it was available without long time for verification by the instrument designer.[30]

During the 1960s and 1970s, NASA's Mars exploration program contributed significantly to the evolution of a global perspective of Earth. This process began in July 1965, when *Mariner 4* became the first spacecraft to successfully reach Mars. Built by NASA's Jet Propulsion Laboratory (JPL), it was the first of three fly-by missions during the 1960s and was followed four years later by *Mariner 6* and *Mariner 7*. These three missions revealed a heavily cratered and lifeless world—nothing like the image of a planet with a thick atmosphere and abundant life that some astronomers and mission planners had expected. In 1971, NASA's Mars exploration effort achieved another first when *Mariner 9* orbited the red planet—no spacecraft in the history of space exploration has more profoundly changed our view of another planet. The orbiter revealed a planet with extreme aerographic features, ranging from giant mountains such as Olympus Mons to fantastically deep canyons such as the Valles Marineris. Many of the images returned to Earth also suggested that water had once flowed on the surface of Mars. This led directly to new scientific questions regarding the history of the red planet. First, what was that world's climate history, and what change had led to the removal of water from its surface? Second, and perhaps more importantly, had Mars ever harbored life?[31]

Carl Sagan postulated that the Martian climate had changed drastically due to variations in the planet's rotation about its axis and orbit around the sun. Because these changes had taken place slowly over time, Sagan believed that only small alterations in the behavior of Mars had led to radically different environmental conditions. Based on this model, it seemed that the red planet could have previously been warm enough and wet enough to support life. Furthermore, because these changes persisted over time, Sagan believed Mars would be able to support life again in the distant future. As a result, the Mars science community largely accepted that microbial life was very likely still present on the planet. In the late 1970s, this belief was tested by the Viking missions, which consisted of two orbiter/lander pairs. For six years, the Viking mission provided incredibly detailed information about Mars. Ultimately, it revealed a cold, dry desert planet with no evidence that life—microbial or otherwise—had ever gained a foothold. Thus, in the span of a decade, Mars went from a living world in pop culture to a dead world (although one that might once have embraced life) in scientific fact.[32]

The effort to better understand Mars enticed scientists to make broad analogies with the planet they thought they did understand better—Earth. Planetary science pioneers were forced to speculate based on very limited data. While the interdisciplinary nature of the new field already troubled many academic traditionalists, this forced speculation challenged "the mechanisms of science, particularly peer review. . . . The controversy remained subdued and relatively minor as long as the planetary scientists only studied other planets—no one else much cared about what planetary scientists had to say about cold, dead Mars. But this changed, when they began to turn their insights, and instruments, on Earth." In 1972, Carl Sagan and George Mullen published an article in *Science* comparing the history of the climates of Mars and Earth. In it, they hypothesized, based upon what had been learned on Mars, that Earth's early atmosphere had been a mixture of carbon dioxide, water vapor, and ammonia. Photosynthesis and the origin of life had added oxygen to the atmosphere, removing the ammonia and cooling the planet.[33]

A British planetary scientist named James Lovelock picked up on this idea, suggesting that the presence of life itself acted as a feedback mechanism with other climate processes to make Earth continually suitable for maintaining life and made Earth different from Mars. In the early 1960s, Lovelock had been a consultant to JPL and had lobbied for planetary missions that would investigate planetary atmospheres for proof of the existence of life (as opposed to planetary surfaces and/or soil). He believed that a planet harboring life would have chemicals in its atmosphere that would not be present on a lifeless planet. Many years later, these ideas about other planets led him to think about the interaction of life with Earth's atmosphere. In 1974, Lovelock wrote a now famous book called *Gaia: A New Look at Life on Earth*, which theorized about how Earth's environment had changed over its three-billion-year history (this book was preceded by a series of articles coauthored by Lynn Margulis, who at the time was married to Carl Sagan). Lovelock and Margulis described the earth's atmosphere as having passed through a transition from a period when it was made up mostly of ammonia and carbon dioxide to one that was oxygen-rich—just as Sagan had done. However, they also suggested that this transition should have destabilized the earth's climate by dramatically reducing the atmosphere's greenhouse capacity. This had not happened. This fact led them to argue for the existence of some kind of active control mechanism of the climate: "They used the metaphor of a planetary engineer, whose employer had assigned him a planet and directed him to maintain a specific set of temperature and acidity specifications for several billion years." The two proposed that this planetary engineer was actually the totality of biologic actors living on the earth and that the natural selection process produced a response to the cooling of the planet that eventually counteracted it. They suggested that biological processes interacted with physical processes during Earth's evolution in such a way as to keep the environment in a state favorable for sustaining life. They conceived of the earth itself as a single organism, a superorganism, which they named Gaia—for the earth goddess of the ancient Greeks.[34]

Lovelock and Margulis were essentially arguing that the earth's climate had been fundamentally altered by the evolution of life. Living things affected the chemistry of the atmosphere, altering its composition. Changing the atmosphere's chemistry affected its radiative characteristics, and over time these interactions produced the earth's current climate. Stated differently, the biosphere itself played a substantial role in making Earth different from its planetary neighbors Venus and Mars. Early life had changed the planet. Even though Lovelock did not consider the impact of modern human activity on the earth, the implications were obvious. By that time, scientists had been able to demonstrate that human emissions of carbon dioxide (a greenhouse gas) were changing the composition of the atmosphere. The logical conclusion, using the Gaia theory, was that humanity had achieved dominance over all other life on the planet and had thus achieved the power to change the earth's climate.

## NASA Wins a Landmark Mission

In the early 1970s, at a time when much of NASA's leadership was still heavily focused on its human spaceflight program, a growing contingent of interest groups was bringing environmental issues to national prominence. In many ways, these issues were enabled by the information provided to scientists and the general public by various weather satellites and Landsat. From an atmospheric perspective, by far the most important environmental issue during this period was ozone depletion. During the previous decade, the Federal Aviation Administration (FAA) had begun leading a supersonic transport (SST) project—an aircraft that was intended to carry 250 passengers, fly at up to Mach 3.0, and have a transatlantic range of 4,000 miles. The SST was still in its earliest development phase when it began to draw the attention of the environmental movement over concerns that SST exhaust would pollute the atmosphere and dilute the ozone layer. In 1971, Congress terminated the SST program before any final determinations could be made about its atmospheric impacts. That same year, Congress established a four-year research effort called the Climatic Impact Assessment Program (CIAP) to investigate the relationship between SSTs and ozone depletion. The primary scientific objective of the CIAP, conducted by the Department of Transportation (DOT), was to determine whether the exhaust of either subsonic or supersonic aircraft would cause ozone depletion. The program was one of several ongoing efforts within the scientific community to better understand the environment in the broadest sense. The CIAP, however, was established with the intention that it would be the leading ozone research program within the government.[35]

While many were focused on SST impacts, NASA had ozone issues of its own. Propulsion engineers were concerned that exhaust from the space shuttle might cause ozone depletion. As a result, NASA Administrator James Fletcher directed the Johnson Space Center (JSC) program office to prepare an environmental impact statement for the launch vehicle. This assessment suddenly made NASA the "five-hundred-pound gorilla of ozone research." In 1973, the JSC study team released a report examining the shuttle's ozone impacts that confirmed the concerns of environmentalists. The space shuttle would release significant amounts of chlorine, a chemical known to destroy ozone, into the atmosphere. By this time, his management team had convinced Fletcher that the CIAP would not be able to effectively answer larger and important questions about stratospheric ozone. The CIAP's four-year timeline inhibited the development of sensors capable of measuring the ozone layer's complex chemistry because they could not be designed, tested, and launched over such a short period. Even with NASA's plans for *Nimbus 7* under way, Fletcher decided to examine whether the agency could do more in terms of environmental and pollution monitoring. In February, Fletcher learned in a briefing that NASA had not yet developed a sufficient number of sensors to detect all of the chemicals associated with the shuttle's ozone problem. Fletcher left the briefing determined to speed the pace of sensor development and implement a new program for stratospheric research after the CIAP ended.[36]

As suggested above, the CIAP's primary scientific objective was to determine whether the exhaust from either subsonic or supersonic aircraft would cause ozone depletion. To answer this question, the DOT initiated an interagency-university-industry study to determine what a future with SSTs would look like (e.g., how many, what kind, and with what emissions) and what that would mean for the ozone layer. The consensus conclusion was that SSTs would have a significantly negative impact on ozone in the atmosphere. However, a controversy arose when researchers at NASA's Langley Research Center suggested that a potential future SST engine would significantly reduce toxic emissions. An executive summary for the CIAP's final report was prepared that based its conclusions on the potential SST rather than the actual SSTs that were under consideration at the time. In January 1975, a *Washington Post* article appeared stating that the SSTs would not have a negative impact on atmospheric ozone despite the fact that most scientists participating in the CIAP had reached a totally different conclusion. NASA was not responsible for this executive summary, nor was the agency implicated in any way in the negative press coverage that followed the report's publication. Instead, DOT officials and the scientists involved in the report's development were subjected to "vituperative attacks on . . . their competence, and their professionalism."[37]

In January 1975, the Senate Committee on Aeronautical and Space Sciences convened hearings to discuss the ozone issue. Committee Chair Frank Moss (D-UT) concluded that the hearings highlighted certain facts about the problem. First, while theoretical projections suggested that the world faced a real crisis, there was little experimental evidence to back up the theory. Second, there were surprisingly few scientists working in the field of upper atmospheric chemistry (perhaps no more than a hundred in the entire world). Finally, efforts to understand upper atmosphere phenomenology had been piecemeal and fragmented. Senator Moss believed the nation needed a major research program examining ozone depletion. Several government agencies seemed to be interested in leading such a program. At the end of the previous year, NOAA's leadership had sent a letter to Congress stating its willingness to take "responsibility to act as lead agency." NASA Administrator James Fletcher wanted an explicit leadership mandate to run a program that would avoid the shortcomings of the CIAP. In essence, this meant the project could not be limited in scope and duration. In contrast to his attitude in 1971, Fletcher was committed to seizing this second opportunity. He created a new Stratosphere Research Program at NASA headquarters to demonstrate that the agency was already organized to lead a new initiative. To the extent that there was competition between NASA and NOAA for this new program, "NASA had the clear advantage in the freedom to maneuver as an independent agency (versus NOAA's subordinate location in the Department of Commerce)." Fletcher and NASA Deputy Administrator George Low were "aggressive in pursuing leadership of the CFC problem during the lengthy ozone hearings" and presented the Stratosphere Research Program as evidence of the agency's commitment. Ultimately, their efforts proved successful. Senator Moss came to believe that the space agency was the right organization to lead the new research effort, given its unique capabilities.[38]

In June 1975, Congress approved the program as part of the FY1976 authorization bill. The legislation directed NASA to "develop and carry out a comprehensive program on research, technology, and monitoring of the phenomena of the upper atmosphere so as to provide for an understanding of and to maintain the chemical and physical integrity of the Earth's upper atmosphere." Any future decision to ban certain ozone-depleting substances would depend on what this research showed. NASA received $7.5 million as a specific budget line-item appropriation for upper atmosphere research. As a result, Fletcher established the Upper Atmosphere Research Program (UARP) within NASA's Office of Space Science and Applications (OSSA). The following year, Congress increased funding for the program to $115.5 million. The large boost was allocated for the development of a new satellite dedicated to ozone research. In 1978, after a short series of design workshops, OSSA began the development of the Upper Atmosphere Research Satellite (UARS).[39]

The decision was made to base the UARS on the *Nimbus 7* platform, but it would be a far more ambitious project. In total, it was expected to include eleven instruments. It would have instrument and science teams focused on developing an improved theoretical and practical understanding of the upper atmosphere. New technology development was projected to cost several hundred million dollars and would take several years. The previous year, Congress had amended the Clean Air Act to require regular reports from NASA on the status of its research. The first assessment was due in 1981, and the agency had been instructed to pay particular attention to ozone depletion. The UARS would be the centerpiece of the space agency's efforts to meet this deadline.

Through the 1960s and into the 1970s, NASA had established its bureaucratic autonomy in human spaceflight and robotic planetary exploration. At the same time, the space agency struggled to define the role of its remote sensing programs. It engaged in a series of turf battles with the Weather Bureau/NOAA, DOD, USGS, and USDA. Through this process, NASA learned that it was often successful in these clashes if it emphasized its experience in researching and demonstrating new technologies. With Nimbus and Landsat, NASA embarked upon a concerted effort to develop new instrumentation and expand laboratory facilities with the goal of positioning the agency to take a leadership role in new research areas in the coming decade. The scientific community was able to utilize Nimbus and Landsat data to dramatically advance scientists' understanding of our home world. The data also made possible the development of an extraordinarily productive interdisciplinary relationship between earth science and planetary science researchers. In the 1970s, NASA leveraged these early successes and pursued a position of leadership with regard to ozone research. In two decades, NASA had learned how to succeed in bureaucratic conflict (with TIROS, ITOS, Nimbus, and Landsat). It had also taken advantage of a policy window opened by the SST debate to gain approval for an entirely new upper atmosphere research project characterized by UARP and UARS.

## CHAPTER TWO

# Evolution of an Idea

On 20 January 1981, Ronald Reagan took the oath of office as the nation's fortieth president. Within a few months, the new president had installed a new leadership team at NASA. For the agency's administrator, he selected former NASA official James Beggs, who had served as associate administrator for advanced research and technology from 1968 to 1969. In this capacity, he had been charged with managing NASA's four research centers—Langley Research Center in Virginia, Ames Research Center in California, Lewis Research Center in Ohio, and the Electronics Research Center in Massachusetts. He had continued his government career after leaving NASA, spending four years as undersecretary of the Department of Transportation. In the mid-1970s, he had returned to the private sector to work for General Dynamics, where he gained a reputation as a strong manager, on programs such as the USAF F-16 fighter jet. He was executive vice president for aerospace programs at General Dynamics when tapped by President Reagan to take over at NASA. He agreed to take the position based on the understanding that the space agency would not be targeted for significant budget cuts by the new administration, even though the month before he was nominated, the White House had produced a report suggesting that it would reduce NASA's budget by $1 billion as part of an overall deficit reduction package.[1]

While serving at NASA during the late 1960s, Beggs had been influential in convincing Hans Mark to become director of Ames. A trained physicist, Mark had been serving as the chair of nuclear engineering at the University of California, Berkeley (and a division chief at the Lawrence Livermore National Laboratory). After an eleven-year tenure at Ames,

Mark had spent four years at the Pentagon—the last two as secretary of the USAF. In this capacity, he had been responsible for a broad range of DOD activities in space. He had also provided critical military support for NASA's Space Shuttle Program, which was then being scrutinized by the Carter administration due to extensive programmatic delays and cost overruns. The space shuttle had been planned as a joint launch vehicle for both NASA and the DOD, but President Jimmy Carter wanted to reduce the number of orbiters and cancel plans for a DOD launch facility at Vandenberg Air Force Base in California. Mark helped rally support within the DOD to keep the Space Shuttle Program on track. Although he was a staunch advocate of increased spending for defense-related space programs, his previous experience at NASA provided him with unique insights into the benefits of the civilian space program (and particularly space science). For this reason, he was a logical choice to serve as NASA deputy administrator under Beggs. At the time of his nomination, it was reported that "he would be the best friend [space science] ever had in that office."[2]

When Beggs and Mark received word that President Reagan planned to nominate them, they immediately met to discuss NASA's future. They agreed on a set of distinct and clear priorities: ensure that the Space Shuttle Program achieved operational status, reemphasize the aeronautics research and space science programs (whose budgets had declined in the face of space shuttle cost overruns), change key personnel, and build political support for a major new initiative. Both men had been at the space agency when a joint station-shuttle idea had been proposed during the post-Apollo planning process only to see the idea degrade to a shuttle-only alternative. Mark had a personal affinity with the idea of a crewed laboratory in space, and Beggs had come to believe that a space station was the only logical choice for a new initiative. In June 1981, both said as much at their confirmation hearings before the U.S. Senate. Beggs believed a space station was needed to ensure continued American leadership in both human spaceflight and cutting-edge technological innovation—which were crucial to promoting national prestige in the ongoing Cold War. As a result, the new administrator was convinced that he would be able to persuade President Reagan to support the station. During their first two years in office, Beggs and Mark worked to set the foundation for this presidential endorsement. At this point, none of the agency's senior leaders were thinking about a major push to add an earth science research initiative to the space agenda. If anything, ongoing remote sensing efforts were at risk under the new administration.[3]

Upon taking office, the Reagan administration began pushing for a wide array of changed national priorities. The civilian space program was not immune to these policy shifts. In 1979, as discussed above, the Carter administration had made privatizing Landsat a key objective. It had also planned to have NASA develop and orbit four additional platforms after *Landsat-3* (with the private sector operating the satellites), ensuring a functioning system until 1994. In 1981, the Reagan administration canceled two of these follow-up missions. It further directed that after the planned launch of *Landsat-5* in 1985, the system would be turned over to private industry. That summer, the Comsat Corporation submitted a proposal to the Department of Commerce suggesting that the company could take over both Landsat and NOAA's polar orbiting satellite system. While that specific proposal was not pursued, the OMB embraced the overarching principle, and NOAA's weather satellites were added to the privatization plans for Landsat.[4]

Environmental satellite development was not the only mission area feeling the pinch of new priorities. In 1979, fiscal constraints had forced NASA to cancel its plans for a mission to rendezvous with Halley's Comet in 1986. This cancellation had nontrivial foreign policy repercussions because the program had been part of a larger international effort to understand the comet. Within the burgeoning planetary science community, the termination increased already growing concerns that robotic solar system exploration had no future. These fears were partially confirmed when the Reagan administration slashed NASA's FY1982 budget $604 million beyond what President Carter had proposed (down to a total budget of $6.1 billion). One of the victims of this reduction was the Venus Orbiting Imaging Radar (VOIR) mission. This left the space agency with only one planetary mission in development, the Galileo mission to Jupiter, and no sense of when any new missions might be approved.[5]

Regardless of any personal assurances that he thought he had received from President Reagan, Beggs would have to deal with even further cuts. In late 1981, the OMB informed the space agency that its FY1983 and FY1984 budgets would be reduced to $6.0 and $5.6 billion, respectively (reductions of 10% and 20% compared with previous planning documents). Beggs believed these large cuts would alter the universe of the possible for NASA and require significant policy shifts. While the OMB favored spreading the cuts across the agency, he proposed canceling the entire solar system exploration and space applications programs and reducing the planned shuttle flight rate. Terminating the space applications program would mean scrapping all weather, space communication, and environmental satellite development programs. NASA plans for the UARS, Earth Radiation Budget Experiment (ERBE), and other post-Nimbus satellites would be put on hold indefinitely. This political ploy led to an intense lobbying effort by the science community and Congress to restore funding for these programs. Beggs ultimately met with President Reagan personally and secured the funds necessary to restore the solar system exploration program and maintain most of the funding for environmental satellite development. Regardless, the reduced funding resulted in significant project delays. For example, NASA was forced to push back the launch of the UARS satellite by four years (from 1985 to 1989).[6]

In March 1982, Burton Edelson joined the space agency as associate administrator for the OSSA. Edelson had been Beggs's roommate during their years at Annapolis. After graduating from the U.S. Naval Academy in 1947, he had served twenty years in the military (including a billet on the White House Space Council staff). After retiring from the U.S. Navy, he returned to school and earned a doctorate in metallurgy from the University of California, San Diego. In 1969, he joined Comsat and rose through the ranks to become director of Comsat Laboratories and eventually senior vice president. After President Reagan was elected, he served as part of the NASA transition team providing personnel and policy recommendations for the incoming administration.

Upon joining the space agency, Edelson immediately pursued a significant change in the relationship between NASA and NOAA. Internal conflicts at NOAA between its operational arm (which maintained satellite operations) and its forecasting arm (which utilized satellite data) with regard to setting requirements for NASA development of new instruments had resulted in many questioning whether continuing the relationship had any value. Tensions between the two agencies had begun to rise when NASA had decided that *Nimbus-7* would be the last of that series. The relationship degraded further as the first generation of NASA and NOAA civil servant leaders began retiring at the end of the 1970s. Matters were not helped by the Reagan administration's budget cuts and increasing space shuttle costs, which meant that all fiscal resources had to be tightly guarded. Shelby Tilford, director of NASA's earth observation programs (and former UARP project manager), recommended terminating the relationship, and Edelson agreed. In 1982, NASA and NOAA formally ended the OSIP; NOAA would not have the funds to support even incremental improvements to its weather satellites, and NASA would no longer consider that task one of its responsibilities. This strain within the space policy community would serve to complicate matters as the scientific community began considering a new space-based earth science program.[7]

## Converging Ideas for Global Change Research

From 1975 to 1981, the National Research Council's Space Studies Board had published a series of reports identifying the primary space science goals for NASA in the coming years. These reports suggested that questions needed to be answered in several key areas, including space and solar physics, astronomy and astrophysics, planetary exploration, comet and meteorite exploration, and the evolution of life in the solar system. The unifying characteristic of these reports was that they involved missions focused on looking outward from Earth rather than using orbit as a location suitable for looking inward toward the planet. In 1982, that changed when NASA tasked the Space Studies Board with conducting an initial study of earth science research priorities. The resulting report, titled *A Strategy for Earth Science from Space in the 1980s, Part 1: Solid Earth and Oceans*, provided a scientific baseline for guiding long-range mission planning and evaluating science content, although it did not identify the specific missions or spacecraft to be developed. It was the first of several reports produced by the space policy community during this period that more fully developed Carl Sagan's idea of using space assets to study Earth from a global perspective. This and other early reports represented an organic outgrowth of the wider scientific community, with NASA contributing and reacting to these developments.[8]

Building on the momentum generated by interdisciplinary studies conducted during the late 1970s, the Space Studies Board endorsed studying Earth using techniques similar to those being employed for investigations of other planets (and providing a similar level of budgetary resources). It argued for a long-term strategy that would begin with missions intended to obtain a global perspective, which would inform the selection of the satellites and sensors needed to conduct more targeted, in-depth studies. The board noted that this approach differed from how past ground-based earth science studies had been conducted, as these previous efforts had used detailed examinations of specific phenomena to gain a global perspective.

Because of the difficulties associated with the prior approach, the report suggested that an understanding of the earth as a global system was "only in its formative stages, and many of the detailed ground, sea, and air observations cannot currently be placed within the scale of planetary-scale processes. The phase of global exploration in the scientific study of the Earth is far from complete." The board wrote, "recent technological developments have made it possible to consider a number of types of measurements from orbiting platforms that address important issues in earth science" and that space measurements could contribute "fundamental improvements in our understanding of the Earth and its dynamic processes." The report reiterated that to fully benefit from the opportunities provided by space-based assets, earth science research should follow the strategy employed within the planetary sciences. It contended that the "investigation of other planets in the solar system has of necessity begun with a global approach, and our knowledge of these bodies has progressed to the point where answers to fundamental questions are being derived from comparative studies. To understand the role of the Earth as a planet in the solar system we must similarly adopt a global framework for its study."[9]

The Space Studies Board identified eleven key areas of study within the earth sciences that would benefit from space-based observation: global atmospheric circulation, atmospheric and climate history, ocean dynamics, atmosphere-ocean interaction, global ice and the hydrological cycle, major chemical cycles (e.g., the $CO_2$ cycle), plate dynamics, evolution of the earth's crust, sedimentary cycles, the earth's internal structure, and the earth's magnetic field. The board included a cautionary note regarding the lack of clear leadership and budgetary support for this massive undertaking:

> Over the period to which this global strategy is directed, we expect that NASA will maintain its role in carrying out supporting research and development programs, especially for the development of flight instrumentation and mission techniques that precede initiation of operational systems. The Committee is seriously concerned, however, that the level of support for and the substance of the instrument development programs at NASA are inadequate to support the strategy and its primary objectives and measurement accuracies contained in this report. The conclusion is not a reflection on NASA's capability; to a large extent it is attributable to the relatively recent beginnings being made by the agency to justify and develop a balanced, long-term remote sensing pro-

gram for earth sciences. Nonetheless, it is a clear indication that a major supporting element of this science strategy, and of the mission planning that grows out of it, needs attention at the earliest opportunity. . . . The overriding implication of adopting this [report's] strategy and its successful implementation is that the United States will embark on a vigorous, systematic program of Earth exploration from space. At present, the lack of a national policy for space exploration is a major obstacle to achieving the present strategy for Earth science.

The Space Studies Board did not endorse any single agency as a lead for this federal effort in earth science, though it did recognize the need for some kind of centralized management, as past efforts had been scattered among several independent agencies. The committee members viewed determining the optimal organizational structure as vital to the success of any future program. Given that NASA would surely be a critical actor in any government effort, the report concluded that there was a need for improved relations between the space agency and the existing earth science community:

> Lastly, during the development of this strategy, the committee became increasingly aware that there is a large segment of the earth-science community that has little or no involvement in earth-science research from space. Undertaking a global program designed to carry out this strategy clearly will require a much larger participation from this community than is currently apparent. There are sufficient precedents in other areas of space science that strongly indicate that close and sustained cooperation between NASA and the scientific community can be achieved. As one means to build to this level of participation, we suggest that NASA demonstrate its intentions through active enlistment and support of available expertise to develop an appropriate experimental and instrumental protocol to address program goals.

In the months and years that would follow, creating such positive and durable relationships would prove challenging, but such efforts were nevertheless central to agency planning.[10]

In February 1982, just before Burt Edelson came to NASA, Hans Mark convened an all-day session with Richard Goody and Michael McElroy.

Goody was a British-born atmospheric physicist who many years earlier, as a graduate student at Cambridge, had constructed an airplane-based infrared spectrometer to measure stratospheric dryness. In the late 1950s, he had joined the faculty at Harvard University and later served as chair of the Space Studies Board. McElroy was an electrical engineer who had spent nearly two decades directing research at Goddard on laser communication systems, tracking and radiometry, and advanced satellite communications technology. He had risen to the post of Goddard's deputy director, where he had become engaged in earth observation work. Years earlier, as the space agency had been working to identify landing sites for the Viking missions, McElroy had been struck by a powerful thought: "You know, we've never done anything like this for the Earth." Now, he and Goody had an opportunity to share these thoughts with Mark. They opened the meeting by expressing their concerns about the "$CO_2$ problem," meaning the impact that a rise in atmospheric $CO_2$ concentrations might have on Earth's climate and its ability to support life. Goody conveyed his belief that these phenomena could best be understood using space-based measuring techniques and remote sensing and proposed that NASA organize a global observation program to study the problem. Mark encouraged Goody to gather leading experts in relevant fields (e.g., meteorology, oceanography, ecology) to determine whether an initiative could be defined in time for UNISPACE '82—a United Nations–sponsored space conference to be held in Vienna that August.[11]

During the final week of June, Goody convened such a meeting at Woods Hole, Massachusetts. The committee that he assembled produced a report that was released publicly the following month. The document opened by suggesting that the "Earth is a planet characterized by change, and has entered a unique epoch when one species, the human race, has achieved the ability to alter its environment on a global scale." As a result, the group recommended the creation of a new program within NASA:

> This report outlines a scientific strategy that would offer a basis for the difficult choices that lie ahead and for the complex decisions that must be made now to protect the integrity of the Earth. NASA could play a central role in this task. The unique perspectives of space observational systems, the ability to manage complex interdisciplinary science programs realized in two decades of planetary exploration, and the overall technical expertise of NASA are essential to the success of the

endeavor. . . . [This] research initiative [would attempt] to document, to understand, and if possible, to predict long-term (5–50 years) global changes that can affect the habitability of the Earth. . . . The program will involve studies of the atmosphere, oceans, land, the cryosphere, and the biosphere. On decadal time scales, these regimes and the cycles of physical and chemical entities through them are coupling into a single interlocking system. Some parts of this system can be studied in a straightforward manner (the atmosphere) and some with great difficulty (the biosphere). A new emphasis for NASA would be to design and carry through studies of the complex interactions. A major effort to involve new scientific talent would be necessary. The program would be large, comparable in size and scope to other major activities of NASA such as the Physics and Astronomy Program and the Planetary Exploration Program. It would require support for at least a decade.

The report identified areas for investigation, including atmospheric processes, ocean-atmosphere interactions, ocean biological processes, and sun-earth interactions. It noted in several places that recent technological advances, particularly in space-based remote sensing, made these new scientific explorations possible. The objective of this new program would be to gain an understanding of the earth's natural cycles and their macrolevel interactions with the specific goal of tying those cycles and interactions to human activity and informing the policy debate regarding the need to enhance planetary habitability. The Global Habitability Report, popularly known as the Goody Report, was the first government-sponsored study to explicitly link scientific research ideas to humankind's interest in maintaining a habitable planet. Further, it laid out a specific plan for obtaining the required measurements using space-based assets to address this new scientific interest.[12]

After the Goody Report was released, Burt Edelson approached the NRC to seek its help in coordinating American scientific efforts on global habitability. He met with Herbert Friedman, chair of the Commission on Physical Sciences, Mathematics, and Resources, to discuss issues associated with establishing such a governmentwide effort. Friedman was "recognized as a pioneer in the space sciences for his contributions to solar physics, aeronomy, and astronomy." From his post at the Naval Research Laboratory, he had been a forerunner in using sounding rockets to conduct solar and atmospheric research. Over the decades, he had become a leading scientific statesman, including publishing an influential book on space science. Friedman informed Edelson of existing plans to hold a workshop in Woods Hole in July 1983 to discuss the idea of a broad international geosphere-biosphere program for research in all areas of earth science. Friedman assured Edelson that global habitability would be part of this program but made it clear that the idea would be broader in scope, encompassing many avenues of research in addition to those associated with space-based observations of the earth.[13]

Over fifty scientists from across the government and academia attended the workshop. In 1957–1958, Friedman had participated in the IGY and believed that this type of international scientific exchange was worth repeating. Twenty-five years after that momentous event, he saw an opportunity to revitalize international discussion of earth sciences. The workshop participants discussed five research areas that offered natural opportunities for coordination: the atmosphere, oceans, lithosphere, biosphere, and solar-terrestrial system. The workshop resulted in the publication of what became known as the Friedman Report, which observed that satellite technologies could revolutionize the study of the planet: "The power of new technologies for remote sensing of atmospheric, geological, biological, and oceanographic conditions promises to revolutionize our grasp of global conditions and our understanding of global change." Like the Goody Report, the Friedman Report recognized the relationship between studying the environment and living in it: "Beyond the intellectual drive to understand basic scientific interrelationships is the practical need to gain a better grasp on how to manage the environment and global life-support systems."[14]

It was Friedman's opinion that the proposed international geosphere-biosphere program could be an efficient structure to coordinate many domestic and international programs over long periods. However, the report did not endorse the wide use of space-based measurements to support the stated scientific objectives, choosing instead to articulate the opportunity for using even simple measurements (e.g., of coastal sea level) to maintain accurate records. Unlike the Goody Report, the Friedman Report did not endorse a new government-led earth sciences initiative. It believed the responsibility lay with the international scientific community. As Friedman would say after the report was published, "Scientists are more than willing to join forces. Governments must be persuaded that it is in their interests to support international cooperation."[15]

The Friedman Report also clearly differed from the Goody Report with regard to its breadth of scientific inquiry. The former report's opening paragraph read, "The concept of an International Geosphere-Biosphere Program, as outlined in this report, calls for [a] bold, 'holistic' venture in organized research—the study of whole systems of interdisciplinary science in an effort to understand global changes in the terrestrial environment and its living systems." The Goody Report was much more limited in scope, focusing primarily on the benefits to be provided by space-based assets. Although Freidman knew that for the international geosphere-biosphere program to make headway, there would need to be "a limited set of high-priority thrusts around which a long-range program can grow," the report did not identify a list of top priorities. It did not endorse, for example, the importance of studying the atmosphere, oceans, and land in order to understand climate change. This meant the report did not favor inquiries focused on short-term timescales (five to fifty years) over those focused on long-term timescales (one hundred to one million years). Finally, the report did not endorse prioritizing research that would provide results within a decade or less. On the contrary, Friedman believed that a well-coordinated international program might not produce its best results for twenty to forty years. These timescales certainly would not provide a strong rationale for government investment in such research projects.[16]

## Global Habitability, System-Z, and the Earth Observing System

Keenly aware of the scientific community's interest in a new earth sciences initiative, NASA began seeking opportunities to respond by creating new programs. In August 1982, James Beggs and Shelby Tilford presented the idea of an international global habitability research program managed by the space agency at the UNISPACE II conference in Vienna. The response was unambiguously negative for several reasons. Many within the American scientific community felt that they had not been appropriately consulted about the contours of such an undertaking. One participant contended that the proposal was "presented without warning [and as a result] it came across like NASA was trying to take over the world." Non-American attendees were "insulted at the implied condescension, and worse, members of the international science bureaucracy saw Global Habitability as undermining . . . existing cooperative programs." Domestic agencies were not pleased because they thought the program was a NASA power grab intended to diminish their ongoing remote sensing activities.[17]

Stunned by this reaction, Edelson approached his government counterparts to "acquaint them with the goals of the Global Habitability program and to encourage their participation." He met with NOAA Deputy Director Anthony J. Calio, USGS Director Dallas Peck, and various other officials from the Department of Agriculture, the National Science Foundation (NSF), and the Department of Education. The results of the meetings were anything but positive. One NASA official was quoted as saying, "The word global habitability became simply a very bad word. The program became very unpopular. So suddenly we had sold a program called Global Habitability and everyone was opposed to it in Washington, except nobody knew what the hell they were really opposed to, because it hadn't been very well defined. And we kept saying, well, yes, but the idea's important. Global things are important." The global habitability initiative quickly faded away. However, as one policy window closed, another one opened.[18]

By this time, the NASA leadership had established a Space Station Task Force to set the foundation for a presidential endorsement of a crewed station program. The group was charged with defining requirements for a space station, analyzing mission concepts and architectures, and addressing management issues. When Burt Edelson came to NASA, one of his first actions was to ask Goddard and JPL whether the space station would have any scientific utility. Both field centers believed that the most likely scientific potential lay in the station's role as an orbital platform for assembling large spacecraft or calibrating instruments before sending them off to explore the solar system. If the space station concept were to include one or more support satellites, however, there would theoretically be some additional science benefits. Goddard and JPL specifically espoused the development of a large polar orbiting platform that would supply a variety of functions, including electrical power, data management, communications, guidance and navigation, stability, and a good view of the entire planet. As Adam Gruen wrote, "Instead of building one satellite for each mission, each satellite requiring its own subsystems, NASA could economize by building one platform to handle all sets of missions."[19]

The Space Station Task Force was directed by John D. Hodge, a British-born aerospace engineer who had been a flight director during Project Gemini and led JSC's Advanced Programs Office for the final three Apollo lunar landings. He had left NASA in 1970 but had returned the previous year when Beggs had asked him to lead the task force. Hodge decided that the group would spend most of its first year concentrating on the missions and requirements for a space station rather than its design. This decision was made both to avoid criticism of early ideas and to establish a foundation for the program, which might ultimately avoid cost overruns in the long term. Hodge worked fast, and by late August, NASA had signed eight contracts for mission analysis studies to "identify and analyze the scientific, commercial, national security and space operational missions that could be most efficiently conducted by a space station." NASA provided these contractors with guiding principles, which included a station that was permanently crewed, shuttle compatible, and capable of autonomous operations. To guarantee the broadest possible political support, the station would be an amalgamation of crewed and robotic elements.[20]

On 31 August, Edelson commenced an internal study examining an "advanced capability Earth remote sensing satellite system for low-Earth orbit." This sixty-day study would provide needed information regarding the feasibility of conducting a scientific study of the earth using a polar orbiting satellite associated with the Space Shuttle and Space Station Programs. In a memorandum providing direction for the study, he referred to this concept as System-Z and provided an indication of the breadth of his interest in an earth remote sensing satellite system:

> Please aim at a satellite system designed for all civilian remote sensing R&D purposes: science—geology, hydrology, ecology, atmospheric chemistry, climatology and oceanography; and potential applications—agricultural assessment, renewable resource monitoring, nonrenewable resource exploration, ocean monitoring, mapping, meteorology, etc. The spacecraft instrumentation should include multi-band optical and IR imaging equipment (MLAs and other advanced scanners), radiometers, spectrometers, and LIDARs. Also, the instrumentation should include microwave equipment (e.g. SAR, scatterometer, altimeter, etc.). The spacecraft should have advanced onboard data processing, data storage, and telecommunications equipment and be designed for read-

out to the ground or to TDRSS, and for store-and-forward operation. The spacecraft should be designed for Shuttle launch and Space Station support. I would like this to be the first spacecraft specifically designed for Space Station operations.

While the relationship between System-Z and the station was not yet clear, Edelson believed that System-Z would be one of the robotic elements that would help the station maintain support within the scientific community. As a result, he believed that the station program would pay for the development of System-Z.[21]

In December, the System-Z concept study team delivered its report to Edelson. It assessed the observational capabilities that could be brought to bear on open questions in the earth sciences as defined by the National Academy of Sciences and the NASA Advisory Committee and in scientific journals. The report grouped the earth sciences into nine areas: atmospheric chemistry, atmospheric circulation, global climate, ocean dynamics, global biogeochemical cycles, global ice and hydrological cycles, continental geology, global biomass dynamics and land cover, and land use dynamics. This list was more expansive than the one initially outlined by Edelson, but it was the study team's belief that these areas of focus appropriately captured both the traditional earth sciences (identified by Edelson) and important emerging interdisciplinary fields. In each area, the study identified three important questions that could best be answered using an earth-observing satellite. Further, the report outlined what suite of instruments could best serve to answer all twenty-seven questions—the results were complicated matrices that populated the briefing materials for the concept study.[22]

The study team also identified a concept of operations for System-Z, including a strategy for system evolution. System-Z satellites would be polar orbiting platforms designed to compose a permanent facility in space. They would be constructed over time in three increments, with each increment adding a suite of sensors to address a collection of scientific issues. The system would utilize the space shuttle for human servicing, integration and testing, and retrieval for upgrade and repair. System-Z would also have several autonomous operational capabilities, including orbital maintenance; diagnosis of in-flight anomalies; acceptance of real-time requests for new observations; and data management, storage, and transmission to the ground. With these capacities, the system would be the first specifically developed to allow for major configuration changes while in orbit. This meshed well with the vision Beggs and Mark had for the space station as a structure that could evolve and grow over time. The concept team wanted System-Z to be the first spacecraft to implement this approach.[23]

That System-Z would be a large, permanent facility had scientific value in and of itself. At a basic level, it would provide the long-term, continuous data collection needed to understand complicated processes such as climate change. Beyond that, however, the grouping of sensors on one large platform allowed for the simultaneous study of related variables such as temperature, humidity, and cloud reflectivity. Collecting this data concurrently would make it easier to understand the complex relationships among these sets of values. Because data would be collected for several variables at once, it could be easily cross-referenced by their space and time relationships. The need for *simultaneity*, for large quantities of data, and for autonomous operation between servicing missions resulted in requirements that were unique and more advanced than anything NASA had previously developed. System-Z would require data processing systems (for transmission and relay) that had not yet been demonstrated, onboard data storage capacity that had not yet been developed, a new approach to software development to handle autonomous operations, and new ground databases capable of accurately cross-referencing and time-phasing data. Finally, the data would be made available to a growing number of researchers in different formats that would meet their needs. The concept team concluded that System-Z was ambitious but "feasible, with exciting potential." The belief was that its primary attributes would enable world leadership in human space operations, autonomous space robotics, sensor development, data processing, database system development, and the earth sciences. In addition, it would provide a new rationale for the human space exploration program. A great deal of additional work was needed to realize these benefits. Specifically, everything had to be examined in much greater detail, and cost estimates (and weight estimates) were required to inform the larger policy debate.[24]

While the System-Z concept study had given Edelson the start he needed, its results were too general to begin any real program planning for polar orbiting spacecraft. Thus, he directed the Goddard to lead a follow-up study to provide a preliminary definition of the scientific requirements and spacecraft alternatives that NASA could begin developing the following year. This new study was to be conducted in close cooperation with the Space Station Task Group, which was also in its earliest formulation phase. To meet these goals, two new internal groups were created. The System-Z Science Group would establish science requirements, and the Technical Advisory Committee would establish spacecraft system architectures, drawing upon senior-level managers from NASA headquarters, Goddard, and JPL, with the latter providing specific support for spacecraft design (particularly the spacecraft bus that would provide power and station-keeping capabilities). Study Director Charles Mackenzie from the Goddard would then integrate the System-Z Science Group and Technical Advisory Committee findings, establish relationships with the various Space Station Task Group study teams, and assist the headquarters leadership team with policy implementation. Building a rapport with the space station groups was considered particularly important because Edelson wanted to ensure that System-Z planning was consistent with emerging space station concepts and architectures. He established a timeline that called for establishing the System-Z Science Group by April 1983, defining the science requirements in a preliminary report by March 1984, and presenting recommendations for the science instruments, mission requirements, and initial spacecraft plans in a final report by October 1985. In the meantime, he needed to dedicate serious effort to finding allies within critical policy communities that would support the ultimate adoption of the program.[25]

As the process of defining the science requirements began, Edelson started lobbying members of the NASA Advisory Council to obtain their support for prioritizing earth science missions. The council provided recommendations to the space agency with regard to programmatic options given specific political and budgetary constraints. The committee was then, and still is, composed of distinguished academics and industry experts in various scientific and technical fields. Edelson interacted regularly with the Space and Earth Science Advisory Committee to discuss future plans. This subcommittee's members had previously expressed concern that the continued deceleration of funding for science initiatives could lead to a logjam of potential new mission starts in the late 1980s. As a result, its objective was to propose mission priorities assuming the space agency continued to face budget pressures or cost overruns. Luckily, Edelson had an important ally on the subcommittee. Francis Bretherton was an English-born applied mathematician who had taught planetary sciences at Johns Hopkins before spending most of the 1970s jointly directing the University Corporation for Atmospheric Research and the National Center for Atmospheric Research.

Although he had since left these administrative posts, he remained highly influential in the field and was an outspoken supporter of giving precedence to NASA's earth science programs.[26]

Bretherton and Edelson both believed the earth sciences needed an implementation strategy, but they differed in their preferred approach to generating this long-term plan. The former wanted the space agency to come up with a plan that the advisory council could review and endorse. The latter wanted the Space and Earth Science Advisory Committee to conduct a more intensive decadal survey that would rank the scientific priorities identified by the Space Studies Board and the Goody Report. This difference in approach was important. Bretherton was not sure the NASA Advisory Council could accomplish this expansive task. "My first reaction was, it can't be done," he recalled later. He was concerned with the disparate priorities within the various earth science disciplines, noting, "They just don't talk to each other." Ultimately, it took a phone call and some personal pleading from Shelby Tilford to convince Bretherton. On 26 May 1983, the advisory council announced the establishment of an Earth System Science Committee (ESSC). Chaired by Bretherton, the new group began work about six months later on what would become a two-year effort to review the science of the earth as an integrated system of interacting components, recommend an implementation strategy for global earth studies, and define NASA's role in such a program.[27]

In April 1983, Edelson reorganized the OSSA. As part of this effort, he put Shelby Tilford in charge of a newly created Earth Science and Applications Division. Tilford was a physical chemist who had become interested in spectrographic research while attending graduate school at Vanderbilt. He had been a research scientist for many years under Herbert Friedman at the Naval Research Laboratory before coming to NASA in the mid-1970s to work on the congressionally mandated UARP. As an experienced program manager and administrator, he was a logical choice to lead the Earth Science and Applications Division. The organization included all of the space agency's environment-related research, which put earth science on an equal footing with the space sciences. This was quite amazing given how little emphasis or funding had previously gone into this area, particularly in comparison to planetary missions. A number of missions were already planned or under consideration. These included the UARS, Ocean Topography Experiment (TOPEX), Geospatial Research Mission (GRM),

Solar Dynamics Observatory (SDO), and Origins of Plasmas in Earth's Neighborhood (OPEN). These missions were all proposed for new starts in the second half of the 1980s. If approved, System-Z would be an addition to this list of potential missions.[28]

By this time, the System-Z Science Group was already up and running. The group was chaired by Dixon Butler, a particle physicist from NASA headquarters, with Richard Hartle, an upper-atmosphere scientist from the GFSC, serving as vice chair. Eighteen respected scientists from academia joined these two to form a twenty-person panel of experts. The group broke into four broad disciplinary areas for consideration: hydrology, biogeochemistry, climatology, and geology. The overarching objectives were the "definition of science requirements for the 1990s; definition of a multidisciplinary approach to requirements implementation; and the development of a straw-man instrument configuration for the 1990s."[29]

The System-Z Science Group quickly reached a consensus regarding what was meant by a multidisciplinary approach to requirements by identifying the synergies enabled by space remote sensing. In a memorandum sent to Tilford, the committee wrote, "There are shared needs to observe the same phenomena for different purposes and there are cases where the same instrument can measure different phenomena in different regions of the globe." In June, the group also agreed to release a statement supporting the agency's other missions, believing these would set an essential foundation for System-Z: "In the discussions of the committee, it is quite clear that the planned missions for the Earth Sciences—UARS, TOPEX, OPEN, and GRM—are all essential to progress in this area of research and that the formulation of System-Z science [requirements] rests on the assumption that each of these missions will be carried out."[30]

Over the next year, the System-Z Science Group and Goddard EOS Project Office worked to define the System-Z science requirements and establish a straw-man instrument configuration (specifically trying to define how the satellites would interface with platforms developed by the Space Station Program). During this period, Edelson was searching for a new name for the polar orbiting satellites—something more appealing than "System-Z." He ultimately chose Eos for the Greek goddess of the dawn. He liked the name for two primary reasons. First, the acronym EOS could be used for a fleet of spacecraft to be called the Earth Observing System. Second, EOS could be used to convey the idea that the program represented the dawn of a new era in earth science and remote sensing.[31]

❧ ❧ ❧

In January 1984, President Reagan used his State of the Union address to announce that NASA would develop a space station. As discussed in the introduction, this decision had come after more than two years of maneuvering by James Beggs. Immediately following the presidential announcement, Edelson established a Space Station Working Group to "identify a combined set of science and application needs for the Space Station." In May 1984, the committee released an interim report that outlined how a station could be utilized for astrophysics, solar system exploration, earth science, life science, and communications—assuming the station provided a minimal set of capabilities outlined in the report. The earth science section included the applications previously identified by the EOS Science Group (no longer called System-Z) and a short discussion of a potential geosynchronous platform to study rapidly changing atmospheric and oceanic phenomena.[32]

During this period, Edelson also engaged the Space Studies Board to "undertake a study to determine the principal scientific issues that the disciplines of space science would face during the period from 1995–2015." This time frame coincided with the planned operational life of the space station. For political and budgetary reasons, Edelson intended to make the station a central component, if not *the* central component, of NASA's future science endeavors. The Space Studies Board created six task groups (earth sciences, planetary and lunar exploration, solar and space physics, astronomy and astrophysics, fundamental physics and chemistry, and life sciences) to pursue these questions and commenced a two-year study to provide answers.[33]

In August 1984, the EOS Science Group space station working group delivered its final recommendations to the Earth Science Applications Division. The panel concluded that EOS's overarching scientific objective should be the development of a detailed understanding of the hydrological, climate, and biogeochemical cycles over both the short and long term. This would be accomplished by taking holistic measurements of the atmosphere, the ocean, and the solid earth. These investigations would require a sustained commitment to continued satellite operation, information system development (needed to collect and store observational data), interaction with the scientific community, and training of future scientific personnel. The study produced an exhaustive list of observational requirements for EOS and additional analysis of the types of instruments that would be needed. Finally, in concert with the Technical Advisory Committee, the group detailed a spacecraft concept that would incorporate all of the needed capabilities.[34]

In total, the EOS Science Group recommended twenty-three instruments to meet the program's observational needs. While the group did not expect that all of these instruments would find their way into space, it hoped that fifteen to twenty might—there would be an ongoing conversation with the project office to determine which instruments would actually fly. There were four general instrument categories: spectrometers and laser instruments used to measure soil moisture content and temperature; radars, altimeters, and scatterometers used to measure ocean dynamics and sea/land ice thickness; various instruments used to measure wind speed and atmospheric content for chemicals such as ozone or hydrocarbons; and assorted instruments to monitor solar activity and measure the earth's magnetic field variations. These instruments fell into three basic technology readiness levels: established (where the relationship between the instrument and a physical parameter was well understood and had been demonstrated), well understood (where the relationship between the instrument and a physical parameter was well understood, but the feasibility of the instrument for use in space operations had not been fully demonstrated), and promising concepts (where it was not clear whether the instrument would provide the needed measurements of physical phenomena). The EOS Science Group suggested a five-year deployment period for an initial set of orbital instruments to be launched, with new instruments added every year during routine servicing as technology advanced.[35]

While much of the EOS Science Group report discussed instrumentation, the committee also highlighted the crucial role played by information systems. It believed scientists needed to have ready access to the data collected by previous missions (e.g., Landsat), missions that would fly before EOS (e.g., UARS), weather satellites, and future EOS instruments. This access would require flexible data systems capable of incorporating and correlating data from EOS and non-EOS platforms to verify measurements that were taken at the same time, thus dealing with the problem of simultaneity. It would also necessitate the development of appropriate software tools and methods for the manipulation, processing, and analysis of the extremely large data sets that would be collected.[36]

In November 1984, building on the recommendations from the EOS Science Group, the Goddard EOS Project Office released an initial report that defined the program as both a measurement system and an information management system. The measurement system was characterized by the spacecraft element—a polar platform to be paid for by the Space Station Program and a collection of EOS spacecraft containing the needed scientific instrument packages. The polar platform would include power systems and spacecraft operational controls for the scientific instruments aboard all EOS spacecraft. In this ambitious proposal, both the polar platform and the individual spacecraft would be serviceable by astronauts during space shuttle missions.[37]

The information management system would be significantly more complex, integrating data at three distinct levels. The primary data would be the observations and measurements collected by the science instruments. The secondary data would be the engineering and reference data collected by the polar platform. Finally, the tertiary data would be collected from other operational satellite systems. All of the data would be collected and managed across several heterogeneous databases that would provide simultaneous access to multiple users. Thus, the information system would have to manage the transmission, relay, and integration of science and engineering data while also documenting the changes and updates made to the databases and maintaining format standards. This was an enormous task, in many respects presenting more significant challenges than the development of the actual spacecraft involved.[38]

Given the scale of the overall program requirements, both mission and science teams would be needed to manage the overall EOS system. The project office recommended that the mission team include a mission operations subteam overseeing day-to-day platform operations and a system support subteam ensuring the functionality of all computer and software facilities. The science team would include a science steering committee (working from headquarters and interacting with the external science community), a science management team (working from the project office and overseeing day-to-day payload operations and scientific plans), and numerous instrument teams (responsible for managing individual instruments).[39]

EOS represented something entirely new for NASA. The scope of its research program, both in terms of the breadth of scientific inquiry and the duration of the investigation, was greater than anything ever undertaken by the space agency. The organization had never attempted to develop such a complex data system. The concept of a serviceable scientific spacecraft (to be maintained and improved over time) had never been demonstrated. As a result, compared with the original System-Z concept, NASA would have much more on its plate. To keep to its schedule, including the release of a request for proposal (RFP) to private industry by early 1986, the EOS Project Office would have to spend an entire year refining as many mission details as possible. Most importantly, it would need a much more rigorously defined cost profile. The question was whether the current administration would support such an undertaking. One thing that would help would be to bring other government partners on board. This would prove incredibly difficult to accomplish.

## EOS and NOAA

In the aftermath of President Reagan's State of the Union address, Edelson had begun briefing government agencies regarding the potential science benefits of the space station. It appeared to him that a policy window had opened to include an earth science component to the overall mission. In May, NOAA Assistant Administrator John McElroy expressed interest in studying how EOS satellites could help his agency meet its operational objectives. He highlighted several important trends in weather satellite development that might lead to corresponding functionalities that would allow both organizations to achieve mission needs. These included the increasing use of sensors for simultaneous measurements of the atmosphere and oceans as well as the growing need for continuous operations. As we have seen, EOS shared these characteristics. Thus, McElroy proposed the immediate establishment of a joint study team to examine commonalities with regard to spacecraft, instruments, and orbits that might serve both agencies. By September, a team was in place and working to determine whether NASA's polar orbiting platform could meet NOAA's mission requirements without significantly altering the expected scientific payoff for EOS. Given that EOS had initially been defined as a stand-alone system supporting the space agency and its research community, the team had to revisit several early plans and assumptions to accommodate a new user and a new agency.[40]

The NASA-NOAA joint study team sought to merge concepts for EOS science spacecraft and NOAA weather satellites. The goal was to determine whether a combined set of spacecraft could accomplish both sets of missions. By 1993, each agency would need to launch two spacecraft to establish an initial operational capability. Ultimately, NASA would need a third spacecraft to accommodate all of its instrument packages. While this overall architecture seemed appealing, there were significant difficulties facing any cooperative plan. First, while NOAA needed once-per-day global coverage (provided by a spacecraft in a high orbit, approximately 850 kilometers), NASA had planned on twice-per-day global coverage (provided by a spacecraft in a lower orbit, approximately 705 kilometers). Second, while both agencies wanted their spacecraft to be serviced in orbit, NASA's plans accepted short-term coverage interruptions as astronauts worked to upgrade the spacecraft or replace instruments. Such a data collection interruption was unacceptable to NOAA, as it would severely hamper its ability to collect weather-relevant information. Third, while NASA had planned to service EOS satellites every twenty-four months, NOAA wanted servicing conducted every twelve months (and wanted servicing to be performed at the satellites' operational altitude, not at a lower altitude safer for shuttle operations). Fourth, while NOAA wanted a direct communications link between its satellites and the ground, NASA had planned to transmit its data via the Tracking and Data Relay Satellite System (TDRSS) and then to ground systems. Lastly, it was far from clear whether the two agencies could merge their ground systems—particularly because NOAA was happy using its existing ground network, while NASA envisioned something brand-new for EOS. Despite these major differences, the NASA-NOAA joint study team recognized that many of the instruments planned for EOS could easily meet specific NOAA needs (and vice versa).[41]

After several months of intense negotiation, the EOS Project Office and the joint NASA-NOAA team reached an agreement. The agencies compromised on a four-satellite fleet (as opposed to five), which would include international and commercial participants. Each spacecraft would be larger than any single NASA-or-NOAA-only spacecraft might have been. NOAA's operational instruments would be spread across two or three satellites, with the same instrument flying on different spacecraft to ensure redundancy. Additionally, some EOS spacecraft would carry one-off NOAA instruments to test future remote sensing objectives. NOAA's instruments would also be coupled with newer NASA research instruments based on the EOS instrument packages defined by the EOS Science Group. If these new instruments proved themselves applicable to NOAA needs, then instruments could be removed in favor of even newer instruments.[42]

The first satellite would have a combined instrument suite, though it would lean heavily toward NOAA's operational mission—it was planned to launch into a high orbit (824 kilometers) no later than November 1993. Over the subsequent two months, two more satellites would be launched. One would be a NASA-only spacecraft, flying at an altitude no higher than 705 kilometers. The other (flying at 824 kilometers) would focus on NOAA's mission but would potentially include international partner instruments from the European Space Agency (ESA) and government of Japan (GOJ) and possibly commercial instruments. This collection of three satellites would constitute the initial operational capability of EOS. Two years after initial operational capability, an additional NASA-only satellite would be launched (flying at an altitude of 750 kilometers). Satellite servicing would be an evolutionary process. Shuttle-based astronauts would service the satellites every two years until the station program had developed an orbital maneuvering vehicle to conduct teleoperation servicing missions. Regarding communications, the joint spacecraft would have both direct-to-ground capability (for NOAA) and a relay capability through TDRSS (for NASA-EOS). NASA-only spacecraft would have only TDRSS capability. The EOS constellation was envisioned to operate for ten to twenty years, depending on the success of the servicing operations and the continuation of the Space Shuttle and Space Station Programs.[43]

By December 1985, NASA and NOAA had reached an agreement to develop, launch, and operate the most ambitious space-based research project ever initiated. There seemed to be almost no downside. The two agencies would be producing four spacecraft, instead of five, to meet their combined requirements—significantly lowering taxpayer costs. Additionally, both would benefit from leveraging the Space Station Program's investment in a polar platform and from relatively inexpensive space shuttle launch costs. The EOS Project Office, working in concert with NOAA, was given the go-ahead to develop an announcement of opportunity to be released in April 1986. Then, the sky came tumbling down.

Within a month, a series of major problems arose that forced an indefinite postponement of the EOS project. First, at a retreat in Bethesda, Maryland, a concern was raised that the joint NASA-NOAA satellite would be too heavy for the space shuttle to lift into such a high polar orbit from Vandenberg Air Force Base in California. As a result, separate launches might be required to get the spacecraft to the appropriate location. The three months required between the two launches had not been accounted for in an already tight schedule. The danger was that the satellite would not come online before the expected failure of NOAA's existing satellite. Second, there were concerns about mission maintenance costs. EOS had been budgeted at $2.1 billion through its initial operational capability, but this had not included the funds necessary for an additional shuttle launch or an extended timeline. Additionally, it was unclear how NASA actually planned to provide the budget resources for the two platforms that would make up its initial portion of EOS. While the Space Station Program had agreed to pay for the first NASA-only satellite, it was not clear how the NASA-NOAA satellite would be funded. The EOS Project Office assumed that the space station program would pay for it, the space station program assumed that the OSSA would contribute, the OSSA assumed that NOAA would contribute, and NOAA assumed that it had no obligations beyond its own instruments. This lack of budget clarity was extremely troubling. Third, there were concerns about the appropriate interface between the platform provided by the station program and the spacecraft carrying the relevant scientific instruments. There were ongoing and seemingly intractable disagreements within the EOS Project Office regarding this key technical issue. Perhaps more importantly, neither the station nor the EOS budget included funding for the orbital maneuvering vehicle that was required to continue servicing EOS. Fourth, there were concerns regarding continued tension between NASA and NOAA regarding the altitude at which the satellites would fly once all three initial satellites were in orbit. NASA was still pushing to lower the first satellite orbit to 705 kilometers, while NOAA wanted an orbit closer to 800 kilometers. Finally, there were concerns about the data processing and ground systems. NASA and NOAA had still only conceptually defined how the data management requirements would be handled. Ideas included operating entirely separate systems, utilizing a separate communications capability on each spacecraft, or establishing a collaborative approach to data management and distribution. Given the importance of the kind of complex information system EOS would need, it was troubling that there was no concrete plan in place for its development.[44]

The cumulative impact of this large number of disagreements had resulted in an agitated scientific community. The members of the EOS Science Group were particularly uneasy. There were growing fears that rising project costs, the inability to finalize technical and funding issues, the time-sensitive nature of NOAA's requirements, and the unknown state of data management systems could lead to a crisis situation. In the worst-case scenario, it was possible that NASA might give the game away because NOAA's immediate needs would take precedence over earth science research interests. At the same time, there were increasing criticisms of the general approach of making the nation's operational environmental satellites dependent on the human spaceflight program. John McElroy said at the time, "Some of my science friends have called me a traitor for even being this positive about space station."[45]

By the end of the month, however, all of the existing concerns about the fate of EOS were moot. On 25 January 1986, the space shuttle *Challenger* was scheduled to launch from the Kennedy Space Center. The mission was carrying the first civilian astronaut in history—a social studies teacher from New Hampshire named Christa McAuliffe. It had been delayed for six days by inclement weather and technical concerns, particularly with regard to the ability of the launcher's solid rocket boosters to operate safely in such cold weather. The vehicle blasted off at 11:39, but seventy-three seconds into the flight, a rubber solid rocket booster O-ring failed and led to a catastrophic failure. The shuttle exploded and crashed into the Atlantic Ocean, killing all seven crew members. This tragedy was devastating for an earth science research program that had been designed around the availability of both the Space Shuttle and Space Station Program. The shuttle fleet was grounded for more than two years as the entire project was reevaluated. This led to a long postponement in the commencement of the station program. Edelson and the rest of those within the space agency trying to create a comprehensive environmental science program had to go back to the drawing board.

## CHAPTER THREE

# A Long Road to a New Mission

The very future of the space agency was thrown into doubt when *Challenger* exploded over the Atlantic Ocean, including any hopes of getting a space-based earth science initiative onto the national agenda. Eighteen months before the tragedy, Congress had passed legislation authorizing the establishment of a commission tasked with defining the long-range needs of the nation relating to the peaceful use of outer space and articulating goals for the civilian space program. Democratic legislators had particularly wanted a more detailed explanation regarding the long-term purpose of the space station initiative. In October, President Reagan had signed Executive Order 12490, which established the National Commission on Space. The following March, he appointed former NASA administrator Thomas Paine to lead the commission. During the remainder of the year, the fifteen-member group held public hearings and began to prepare its report.[1]

In December 1985, Jim Beggs had been forced to take a leave of absence to fight fraud charges stemming from his tenure at General Dynamics (he was later exonerated). Only one week earlier, Hans Mark had stepped down from his post. William Graham, a California Institute of Technology (Caltech)– and Stanford-trained physicist, replaced him. Graham had spent his early career working for the Air Force Weapons Laboratory and the RAND Corporation before founding his own consulting firm. For the previous three years, he had served President Reagan as chairman of the General Advisory Committee on Arms Control and Disarmament. He became acting NASA administrator at a critical time. The National Commission on Space was preparing its final report, the space station program was preparing its first official budget, and the Goddard was pushing its concept for an EOS program. The situation became far more complicated, however, when the shuttle disaster occurred the following month.[2]

President Reagan reacted quickly, appointing a commission chaired by former Secretary of State William Rogers to investigate the accident. Although Graham had ably weathered the storm during the months after the tragedy, President Reagan decided to appoint James Fletcher as the space agency's next administrator. Fletcher was a distinguished choice. He was a Caltech-trained physicist with an eclectic background, having held teaching positions at Harvard and Princeton, worked for nearly two decades in industry, and served for seven years as president of the University of Utah. In 1971, he had been appointed by President Nixon as the fourth administrator of NASA during the period when Project Apollo was coming to an end—he had overseen the Skylab and Apollo-Soyuz missions and approved the Voyager program. Perhaps most importantly, he had inaugurated the Space Shuttle Program during his tenure. After leaving the space agency, he had joined the faculty at the University of Pittsburgh but had remained active in national policy. When Fletcher began his second stint leading the space agency, NASA was under siege for the management and technical decisions that had led to the *Challenger* accident. In June, the Rogers Commission concluded that the accident had been caused by the failure of an O-ring seal in the shuttle's right solid rocket booster at liftoff. This was the result of a design flaw in the booster and the agency leadership's decision to override technical concerns raised by Morton Thiokol, the company that had built the vehicle, about launching it in cold weather.[3]

In the interim, the National Commission on Space had published its findings in a glossy volume titled *Pioneering the Space Frontier.* The report stated that the purpose of the space program was the "development of the space frontier, advancing science, technology, and enterprise, and building institutions and systems that make accessible vast new resources and support human settlements beyond Earth orbit, from the highlands of the Moon to the plains of Mars." The group outlined a plan that would establish lunar outposts by 2005 and colonies on Mars by 2035. Administrator Fletcher called the report "an exciting and challenging blueprint for the future [that sets a] bold course for our nation [and] demonstrates the need for the space station." While the task force had been tasked with developing such an ambitious plan, many members of Congress found the report too bold to be politically realistic in the wake of the *Challenger* disaster. Some critics referred to it as a "wish list for space cadets."[4]

Unfortunately, the vast majority of observers focused on the report's human spaceflight recommendations and largely ignored its science proposals. Most importantly, the commission had essentially endorsed the EOS concept, saying, "We propose that a long-range global study of the Earth be taken . . . with particular attention to processes that affect, or are affected by, human activities. A vigorous and systematic study of the [Earth] from the Space Station complex in low-Earth orbit will be an essential component of the project." Under different circumstances, this might have been a great victory for proponents of such a mission. However, the space policy community mostly ignored the recommendation.[5]

In August, NASA suffered another setback when a small Aries rocket carrying an astronomical research instrument exploded shortly after takeoff. Not including *Challenger*, it was the fifth rocket failure that year. As a result, the entire space program was facing a crisis of confidence. Press reports cited anonymous government officials and industry leaders stating that the United States lacked any clear direction in this arena. They argued that there was no clear policy for responding to the launch failures or to provide NASA with guidance for the future. Furthermore, there was concern that the Reagan administration was disinclined to provide the needed direction. The harshest criticism came from those who said the Soviet Union had taken the lead in space exploration. Fletcher had to acknowledge that this was true:

> With Mir they already have a fairly good station and the ability to add to it. They can put together a space station now not too different from what we are going to have in 1994. It's quite clear that by the time we get through with our space station, they may be on their way to the next step. [We] can't just sit here and let it happen. We have to decide where we want to lead and where we are willing to relinquish that leadership to someone else. We need to start looking at the next five-to-ten years to get set where NASA ought to be in relation to the rest of the world.

To accomplish that goal, Fletcher established new teams at NASA headquarters to develop a leadership strategy for the space agency. He directed each of his associate administrators to identify the top priorities that would rely on or benefit from the Space Shuttle and Space Station Programs as well as an aerospace plane, a single-stage-to-orbit vehicle concept that was supported by the Reagan administration. He established an Advanced Program Studies office, headed by Ivan Bekey from the Office of Space Flight. This office was intended to evaluate the advanced technologies required to carry out the ambitious programs in *Pioneering the Space Frontier*. Finally, he asked Sally Ride (the first American woman to fly in space) to produce a report that coordinated the output of these studies with recommendations provided by the NASA Advisory Council. The big question for those interested in commencing an orbital environmental research program was whether they would gain support from any of these studies or whether the entire notion would fade into the background.[6]

## Second-Generation Global Change Studies

While these larger events were shaking the American space program, Francis Bretherton's ESSC was finally completing its two-year effort to define NASA's research goals in this field. In June 1986, the study team's recommendations were rolled out at a joint press conference sponsored by NASA, NOAA, and the NSF. Bretherton noted during the well-crafted event, "From the outset, we realized that we had to look at NASA's role in a broader context than just NASA programs. NASA wasn't the only, or even largest, agency looking at the Earth. So we set up a liaison program [with other agencies]." Having reached out to NOAA, the NSF, the USGS, and the White House OSTP, the report took into account a broad range of agency interests regarding remote sensing and earth science. As a result, "one of the striking things about [the report] was the lack of any major turf fights." Importantly, the study was the first to equate the term "earth system science" with global change research. One of the committee's significant accomplishments was to identify "a unifying scientific strategy by confirming that the concept of the Earth as an integrated system indeed provides strategic guidance for future Earth-science investigations."[7]

A good deal of the report was dedicated to a comprehensive overview of the earth sciences, including the solid earth and plate tectonics, the oceans, the biosphere, and the biogeochemical cycles. The study team stated that the "central approach of Earth System Science is to divide the study of the Earth by timescales, not disciplines. . . . Through this central approach, we thus seek to (1) describe, (2) understand, (3) simulate, and (4) predict the past and future evolution of the Earth on a planetary scale."

The task force categorized description and understanding as the "Goal of Earth System Science" and simulation and prediction as the "Challenge to Earth System Science." The Bretherton committee identified a two-phase strategy for earth system science. During the 1990s, the first phase would require a commitment to continuing already planned operational and research satellite programs at NASA and NOAA in concert with in situ research programs at the NSF. The foundation of this stage would be NASA research platforms such as the ERBE satellite, UARS, TOPEX/Poseidon (with France's Centre National d'Études Spatiales), and Geopotential Research Mission. After space station operations began, the second phase would include a fully developed earth observation system, other specialized research missions, and advanced numerical modeling to enable predictions. The report outlined specific roles for NASA, NOAA, and the NSF in each phase. It would be NASA's job to develop the needed technologies and advance key instruments and computational systems. NOAA would be tasked with maintaining all operational elements of an integrated satellite system as well as long-term data sets. Finally, the NSF would be charged with funding basic research (particularly in support of in situ measurements) and coordinating international participation. In identifying a role for each agency, the Bretherton committee effectively endorsed the plans each had for conducting global change research. This wasn't the end of the line for the ESSC, as it was only the first part of its effort to define NASA's earth science research goals, but it would take two more years of hard work to complete the task that it had begun in December 1983.[8]

That same summer, the EOS Science Group produced a second report that significantly expanded on its effort from two years earlier. Titled *From Pattern to Process: The Strategy of the Earth Observing System*, the document provided a detailed explanation of how the space agency would manage and implement its EOS mission. The report initially reviewed the project's core instruments, discussed the proposed sensors, and provided the basic parameters for the EOS Data and Information System (EOSDIS). It also identified the NASA centers that would be responsible for developing each instrument. The bulk of the monograph outlined what specific science questions were of greatest interest to earth scientists. Furthermore, it described how EOS instruments and science investigations in solid earth processes, hydrological dynamics, atmospheric processes, and biology would contribute to the resolution of these knowledge gaps. The committee outlined seventeen specific investigations by discipline (with several questions associated with each) and over a dozen questions that were multidisciplinary in nature (e.g., questions about air-sea interactions, sea-ice dynamics, and coastal dynamics). *Patterns to Process* served as NASA's tactical complement to the cross-agency strategy of the Bretherton Report.[9]

The final global change study submitted during this period was from the Committee for an International Geosphere-Biosphere Program. Jack Eddy, a solar physicist and astronomer who had conducted groundbreaking research on the Maunder and Spörer Minimums, chaired the group. The NRC formed the group to "give the program a clearer focus, by selecting a cohesive set of compelling problems that clearly require an international effort—questions for which we now have adequate technology and a sound scientific base, which truly need a new program for successful resolution." The study team cited a growing interest within the scientific community in understanding the earth, writing that such an understanding would require "a global view and a new effort to study the Earth and its living inhabitants as a tightly connected system of interacting parts." The Eddy Report identified a focusing objective for the undertaking: "To describe and understand the physical, chemical, and biological processes that regulate the Earth's unique environment for life, the changes that are occurring in this system, and the manner in which they are influenced by human actions."[10]

The Eddy Report dedicated an entire chapter to the value of understanding global change from the perspective of how it was affecting human life. The committee cited the need for improved modeling and simulation techniques to understand these impacts. Noting that a program of global change research would "lean on spaceborne observations for a global perspective," the monograph made it clear that it was not endorsing "a space program, [because] the preponderance of needed science would deal with the processes of change, and the vast majority of measurements would need to be made on the ground and on the oceans, from within the habitat of life." This was a blow to NASA's ongoing efforts to gain approval for EOS, although the impact was somewhat minimized because the committee had decided not to endorse any specific programs. Shortly after the report was made public, the NRC established a permanent Committee on Global Change. John Eddy was soon hard at work in the international and domestic communities trying to establish priorities for cooperative scientific studies.[11]

## A New Ozone Debate

Despite the enthusiasm of the progenitors of the global change movement, it was far from clear whether there was any actual political support for the programs advocated within the Bretherton and Eddy Reports. There were serious doubts within bureaucratic circles and the scientific community regarding the backing to be expected from President Reagan or Congress. It was unclear whether NASA, NOAA, and the NSF could effectively work together to communicate their collective interests to national decision-makers. These agencies had been brought together to tackle the need for better information about the state of global change, but a tight fiscal environment and potentially hostile Reagan administration characterized the political landscape. As a result, it proved very difficult to gain approval for a concerted government program to study global change—even though the scope of the problem was better defined by the mid-1980s. What was needed was a policy window that would provide an opportunity to gain political support for the required research program. Luckily, just such an event was in the offing.

In 1985, a British research group working in Antarctica published an article in *Nature* declaring that there had been a 40 percent loss of ozone in the atmosphere above that southern continent. To stratospheric scientists, these results were nothing short of shocking—particularly because the paper did not explain the mechanism that had produced this significant effect. What was most surprising to NASA scientists, however, was the failure of the *Nimbus 7* spacecraft to find such a hole in the ozone layer. *Nimbus 7* was tasked with looking for damage to the atmosphere, including the ozone layer. After a long examination by scientists at the Goddard, it was discovered that a flaw in their software design produced images that ignored the Antarctic ozone hole. Once the software flaw was fixed, the satellite began providing visual proof that there was an *ozone hole* the size of the continental United States in the sky above the icy landmass.[12]

In June 1986, the Senate Subcommittee on Environmental Pollution chaired by John Chafee (R-RI) held a hearing to review "the status of environmental and climatic research on global temperature increases, and assess the outlook for continued depletion of atmospheric ozone." The committee called NASA's Robert Watson, a British-born chemist who had been leading a periodic assessment of stratospheric ozone that was mandated by the National Climate Act of 1978. It also called William Graham, who was about to leave NASA to begin working as President Reagan's science adviser. Finally, it called James Hansen, the director of NASA's Goddard Institute for Space Studies in Cambridge, Massachusetts. Hansen would ultimately become one of the most important and highly controversial figures in the development of global change research. After obtaining degrees in physics and astronomy, he had gone to work at the Goddard Institute and conducted groundbreaking research on the Venusian atmosphere. By the early 1980s, he had begun applying the models that he had developed to understand our sister planet to better describe what was happening in Earth's atmosphere—specifically examining the effect of trace gases and aerosols on the planetary climate.[13]

On the first day of the hearing, Watson and Hansen captured the attention of the committee and the media when the discussion turned to the impact of environmental changes observed by the scientific community. Discussing the ozone layer, Watson showed a dramatic movie of *Nimbus 7*–derived images that showed the growth of the ozone hole above Antarctica. The panelists highlighted the potential negative impacts for animal, plant, and ocean life on the planet. Perhaps more importantly, they discussed public health concerns related to an increased risk of humans developing skin cancer. Hansen discussed how the buildup of greenhouse gases (e.g., carbon dioxide, methane, nitrous oxide) in the atmosphere would lead to increased global temperatures early in the next century. He said these temperatures would be "well above any level experienced in the past 100,000 years." He added that there was already a broad consensus within the scientific community that the greenhouse effect was accelerating increases in global temperature. Watson concurred, saying, "I believe global warming is inevitable. It's only a question of magnitude and time."[14]

When asked about measures that might be implemented to mitigate the potentially negative effects identified by the panel, Watson argued that there was a need to control the emission of chlorofluorocarbons and excess nitrogen. He suggested that while reducing $CO_2$ emissions would also be beneficial, it was hard to assess how best to accomplish that goal. Hansen's reply, however, probably did more to confuse the issue. He began by suggesting that the scientific data provided conclusive evidence that the earth was warming and would continue to do so unabated unless action was taken. But when he came to discuss solutions, he backed off somewhat:

> I'm sorry if I sound like a befuddled scientist, but I would like to understand the problem better before I order dramatic actions. What I'd

> like to see, most of all, is global observations during the next decade, observations of the atmosphere, the oceans, and the land surface, which allow us to see what is happening better and allow us to develop and test the models to represent what is happening.[15]

While this answer may have been appropriate when speaking at a scientific meeting, it lacked the authoritativeness and urgency necessary to sway national political leaders to take swift action.

William Graham testified later that day before the subcommittee. While he was there as NASA's deputy administrator, he was asked questions and gave responses as if he were already President Reagan's science adviser. He agreed that there seemed to be a global warming trend but suggested that the scientific data was not conclusive enough to take immediate action. His assessment was as follows:

> There is no doubt we are moving toward a warming trend. I think that has been well established, and that is a matter of considerable concern, and we have to know the full implications of that to the climate. I am not prepared today to state what specific action should be taken to accommodate that, but there is no question that if the present research is borne out by continued observation and continued scientific exploration that we should consider taking further action.

This wait-and-see approach essentially endorsed Dr. Hansen's earlier statement. Graham praised the Bretherton Report as the appropriate road map for this continued scientific exploration—with UARS and EOS being critically important. He concluded, "It will be several years before we have a comprehensive understanding of the global climate, and that could even be measured in decades. I think that our understanding of ozone may come more quickly, though it has defied a complete modeling to this point."[16]

The result of these Senate hearings was predictable. While there was some political consensus that the nation should move forward to deal with problems related to the ozone hole, the opportunity to take early action on global climate change had been lost. After the hearings, Senator Chafee called on President Reagan to take up the issue of global warming at the next international economic summit and suggested that he would propose a ban on the use of chlorofluorocarbons in air conditioners. While little action was taken on the larger climate change problem, this hearing did eventually lead to the American ratification of the Montreal Protocol, which banned chlorofluorocarbons. A few weeks after the hearings, Senator Patrick Leahy (D-VT) followed up specifically with NASA Administrator Fletcher. He expressed his interest in a coordinated NASA, NOAA, and NSF research program on global climate change. His letter explicitly referenced the Bretherton Report and expressed a desire to see "an aggressive research program to ensure that our decision makers have the information that they need to develop timely policies to protect the planet." There would be a long wait for such a program to emerge, providing space for national leaders to ignore the rising alarm among climate scientists.[17]

## Battle to Control Global Change Policy

By this time, ideas for a coordinated federal effort focused on climate change had been kicked around the Executive Office of the President for quite some time. Richard Johnson, OSTP's associate director for space, had been working to craft a national global change research policy for nearly two years. As a space scientist, Johnson was convinced that NASA had to be a key player in such a program. A White House aide recalled a few years later that after the Senate hearings, President Reagan basically asked "[whether there is] anything to this climate change issue, and if there is, what am I, as President of the United States, supposed to do about it?" The immediate response was to create a working group on climate change within the White House Domestic Policy Council to answer this question. In late summer, NOAA Administrator Anthony Calio was tapped to lead the working group. Calio had grown up in Philadelphia before studying physics at the University of Pennsylvania, Stanford, and the Carnegie Institute of Technology. He had joined NASA in 1963 and worked his way through the ranks to a senior leadership position before leaving the agency to take the reins at NOAA in 1985.[18]

The working group did not actually accomplish much except to recommend the creation of a more permanent (and potentially more influential) version of the working group. Calio believed NOAA's role in global change research needed to be bigger than that envisioned by the Bretherton Report. "If we didn't get on board as a visible, high profile player," he said later, "NASA and NSF would run off with the program." Calio was highly ambitious. When President Reagan had taken office and Calio was serving as an associate administrator, he had made a play for the top job within the space agency. He recalled later that he had badly wanted that position but lost out to Jim Beggs. Now, as the head of a rival bureau, Calio believed that "global change might be his own ticket to the big leagues." He saw an opportunity to nullify any advantage NASA might have by elevating the climate change working group and controlling federal global change policy from the White House—with him running the show.[19]

On 1 October 1986, William Graham was officially appointed as presidential science adviser. He came into office planning to change the way OSTP operated, believing that the organization had little standing within the bureaucracy and therefore could not bring senior officials to the table when negotiating interagency initiatives. As a result, OSTP was out of the loop while the real decisions were made elsewhere. Graham had strong opinions about the government's responsibilities to carry out earth science research and particularly of NASA's role in this process. He later recalled,

> There were instances where the NASA group would tour the South Pole, make a bunch of measurements, come back, step off the airplane, and give a press conference at the bottom of the steps. I remember one of them that had a statement like this: "We're getting different results from those we predicted; therefore the situation must be worse than we thought." Now, this is not good science, not what you'd call a thoughtful peer review process. This is science by press conference. It has proven to be a relatively powerful way of generating Congressional support for these programs, and scientists are certainly very interested in that, but it's neither good science nor responsible supervision of the work. But the alternative they give you is to have a headline that says, "NASA leadership suppresses concerns of the staff," and that's worse. So it seemed to me that we needed a body that in addition to coordinating and preventing duplication would generate cross-checks on the work: reviews, imagination, and sources of ideas on the problem of the climate.

This attitude matched the prevailing opinion within the Reagan administration that Robert Watson and James Hansen had too much of a "sky is falling" attitude. There were also significant concerns within the White House OMB that an overreaction to the global warming issue could devastate the American economy.[20]

The budget office was equally concerned that the combination of NASA's EOS, NOAA's Climate and Global Change Program, and the

NSF's Global Geosciences Program would break the federal budget. Jack Fellows, an examiner within the OMB's Science Branch, had calculated that carrying out the goals set forth in the Bretherton Report for each of these agencies would cost over $1 billion. He was floored. He later recalled thinking this money

> could probably be spent in a better fashion than it had been. They were doing excellent research, and their research reflected their agencies' missions, and nobody wanted to take anything away from that, but you could add some marginal money to focus all that talent, all those resources that were going to be spent anyway, but have a focus, a target. . . . It seemed to me that if we wanted to do something in this area nationally, then we all had to speak the same language and be going in the same direction.

Fellows had been involved in the informal discussions that Richard Johnson had organized before the Domestic Policy Committee working group was created, so he knew all of the players. As a scientist himself, and not just a budget wonk, he had the advantage of being capable of discussing the global change research goals with the various agencies without being out of his depth in terms of the technical details. This also provided him with leverage to lobby for controlling federal global change policy from the White House, which made him an ally of Calio.[21]

On 6 March 1987, William Graham established just such an organization when he formed the Committee on Earth Sciences (CES) within the Federal Coordinating Council on Science, Engineering, and Technology. This committee was established to "address significant national policy matters which cut across agency boundaries; review federal research and development programs in the Earth sciences; improve planning, coordination, and communication among federal agencies engaged in Earth science research and development; and, develop and update long-range plans for the overall federal research and development effort in the Earth sciences." Membership on the CES was fairly broad, including representatives from the Departments of Agriculture, Defense, Energy, the Interior, and State as well as the Environmental Protection Agency (EPA), NSF, NASA, NOAA, OMB, OSTP, and Council on Environmental Quality. Not surprisingly, Anthony Calio (chair) and Richard Johnson (vice chair) were selected to lead the new committee.[22]

The first CES meeting was held the following month. During the meeting, Calio suggested that he wanted the committee to be the focal point for both science and policy coordination. Furthermore, he believed the organization should be the venue for agencies to come to an agreement regarding the future direction of government-funded earth science research, utilizing a permanent secretariat paid for and staffed by the agencies represented on the committee. Unfortunately for Calio, the members of the committee flatly rejected this structure. NOAA's Eileen Shea later recalled, "Virtually no part of Calio's agenda failed to excite major opposition from some, in some instances all, quarters. There was general interest from the agency representatives in learning what each other was up to, and at least a theoretical commitment to coordination, but nobody was in the least willing to sign on to a process and subordinate themselves to a committee structure they understood not at all." NSF Director Erich Bloch objected most strenuously, as he believed the new committee would subsume a role that had generally been performed by his own agency. NASA was not interested in abdicating any of its decisionmaking authority about satellite and instrument development to a committee essentially run by NOAA. Most surprisingly, given the earlier role played by Jack Fellows, the OMB was also opposed to this plan. David Gibbons, who was representing the budget office, made it clear that determining what research would get funded would continue to be a decision made during the normal fall budget process. OMB was not going to simply accept the priorities set by this upstart committee and its leadership. "The first meeting was a disaster," recalled Shea. Within months, Calio had resigned as NOAA administrator and left government service. It took several months for NASA, NOAA, and the NSF to repair the interagency process that had been strained under the auspices of the CES. In the meantime, the space agency decided to push ahead with its own agenda.[23]

## NASA Moves Forward

While NASA was struggling to maintain its authority in setting global change research policy, a much greater worry for the space agency was uncertainty about the future of EOS. The *Challenger* disaster had created serious concerns about national space launch policy, including whether the orbiter would continue to be the preferred vehicle for science payloads (including EOS). On 15 August 1986, President Reagan had issued a statement articulating a new direction for the Space Shuttle Program. He had announced that the nation would construct a new orbiter to replace the *Challenger* but that commercial satellites would no longer be launched on the shuttle. In December of the same year, this initial statement was followed by National Security Decision Directive 254, which promulgated a new American space launch strategy. The administration concluded that "NASA will use the Shuttle where the unique capabilities of the STS [Space Transportation System] are required to support civil research and development programs." The space agency was given express permission to use other launch vehicles, but it would have to contract with industry providers for the use of existing launch vehicles (rather than obtaining authorization to develop a new expendable launch vehicle). EOS had initially been developed with the idea that it would rely upon the shuttle for launch and astronaut servicing, but the big question now was whether it would be better to develop an alternative approach.[24]

For a year, the EOS program had been in stasis while NASA's leadership addressed broader problems. This had delayed the release of the announcement of opportunity that would have started the competition for funding to develop instruments or experiments for the platforms. Finally, on 15 December 1986, Administrator Fletcher provided the project with clear directions. The program managers were told to make changes to significantly reduce the EOS budget while minimizing any negative impacts on the science mission as much as possible. This decision was driven by Fletcher's desire to sustain the current momentum associated with the project. No conclusion was reached regarding the vehicle that would launch the EOS platforms into orbit. This was worrisome for many within the larger policy community who were trying to make their own plans for the future.[25]

In March 1987, the space agency received a letter from Thomas Pyke, who had replaced John McElroy as NOAA assistant administrator for the National Environmental Satellite, Data, and Information Service. The communication expressed Pyke's concern that any further delays in EOS planning could jeopardize the nation's ability to maintain continuous global coverage with polar orbiting weather satellites. He asked for assurances that NOAA's operational mission would continue to be a priority for the space agency. He also wanted to ensure that agency leaders understood the time-critical nature of the EOS procurement for NOAA. Finally, he wanted to know what options NASA was considering as an alternative to EOS. Two months later, Shelby Tilford sent a reply to assure Pyke that NASA's plans for the polar platform would bring EOS online by October

1995, which would meet NOAA's need. He additionally conveyed the news that NASA had finally released an RFP for three satellites that would serve the nation in the interim. Tilford, however, was honest in his recognition of potential uncertainties that EOS might face: "I believe it would be prudent for you to budget for at least one additional 'free flyer' as a backup." While this letter was meant to be reassuring, it did not alleviate the concerns of NOAA's leadership team.[26]

In July, Burt Edelson retired from NASA to accept a position at The Johns Hopkins University Foreign Policy Institute. His successor was Lennard Fisk, a professor of physics from the University of New Hampshire. From 1971 to 1977, Fisk had worked at the Goddard as an astrophysicist. He had also been a member of the Bretherton Committee, so he was intimately familiar with space agency plans in relation to the earth sciences. As he said at the time, he was not taking the job to "preside over the demise of space science." Fisk believed that NASA had to build momentum for the space sciences by initiating a series of new projects (one per year), starting with EOS and a new planetary mission. His objective was to develop a strategic plan for where the science program should be by the mid-1990s and then set out to make the contemporary moves necessary to establish that program.[27]

As the EOS program office began this planning process, it used the Bretherton Report as a road map. The project goal was to describe, understand, simulate, and predict the past and future evolution of the earth on a planetary scale. NASA also remained committed to the idea of simultaneity, which drove the concept of four instrument packages launched over time to complete the research program's space segment. Notably, Fletcher's direction to reduce the program's budget profile did not result in significant changes to the *content* of EOS, just a reduction in the cost estimates presented. On 8 July, at a meeting with the administrator, the EOS team raised two critical questions. First, what launch vehicle would be used to place the platforms into polar orbit? Second, would astronauts or robotic spacecraft service these spacecraft?[28]

The space shuttle could place only 14,000 pounds into the polar orbit planned for EOS at an altitude of 824 kilometers. This meant that the science payload could be only 300 pounds after accounting for the weight of the polar platform (the spacecraft bus that would provide power and communication), onboard propellant (necessary because the shuttle needed more fuel than it could carry to access the polar orbit altitudes needed for

EOS), and structural support. The EOS team had designed a 7,716-pound science payload. Put simply, the shuttle was not capable of orbiting the planned EOS platform. This realization was not new. The leadership of the EOS and Space Station Program had known about the issue for more than eighteen months. The original solution to the problem was to use an orbital maneuvering vehicle that many assumed would be part of the space station. This vehicle, however, had not been included in the station plans. As a result, the orbital maneuvering vehicle solution was nullified. Because no one on the EOS team wanted to consider a smaller satellite, the team recommended launching the entire space segment aboard a USAF Titan IV. This vehicle was capable of launching the entire polar platform–EOS spacecraft combination with more than a 1,000-pound safety margin.[29]

If the shuttle were not used to launch the polar platform, however, orbital servicing would likely not be possible. Even if launched by a Titan IV, the shuttle would not be able to reach the EOS orbit unless it were launched from Vandenberg Air Force Base in California. The problem was that those launch facilities had been mothballed after the Shuttle *Challenger* accident, and it seemed highly unlikely that this decision would be reversed. There were also concerns about the impact on astronauts of the more dangerous radiation environment in polar orbit, which meant that any servicing would have to be conducted by a combination of extravehicular activity and robotic operations. As a result of these issues, the EOS team recommended the creation of a joint EOS–space station study to investigate servicing options. The goal was to compare shuttle-based servicing versus robotic-only servicing (which would presumably use an expendable launch vehicle to place the robotic servicing spacecraft into the proper orbit). The analysis would include an assessment of the cost profiles, technical performance, and impact on the science mission for each alternative.[30]

At the end of the meeting, Fletcher agreed that the Titan IV should be used as the baseline vehicle for designing EOS platforms. He wanted the program to assess the potential, however, of using the smaller and cheaper Titan III or Delta II launch vehicles, determining what design changes would be required and whether these modifications would adversely impact the science mission. Fletcher decided that the station program would continue developing the polar platform so that it would still be compatible with the shuttle, making it useful for other free-flying spacecraft that might come along during the station era. Finally, he agreed to establish the servicing study and added another requirement for the analysis: He wanted the joint study team to consider a change in the EOS parameters. Instead of assuming that the spacecraft would meet their intended fifteen-year operational life via servicing upgrades, he wanted the team to consider replacing or upgrading the satellites in their entirety without servicing. A week later, the space science and space station offices cosigned a memorandum charging the director of the Goddard to lead this study. It would take nearly a year to complete the review.

In August, Sally Ride's team completed its review of national space policy—which had been ordered after the National Commission on Space submitted its report. For nearly a year, Ride had worked with a small group to develop a new set of missions and objectives for the space agency. The goal was to reestablish NASA's leadership position in space and move the organization forward in the aftermath of the *Challenger* accident. The Ride Report did not set out to reinvent the wheel, as she would later recall:

> We started by reviewing all the studies that had been done, either by NASA or for NASA, over the previous ten years. We wanted to see what was lying on the shelf already, what recommendations had been made, what consistencies there were in the recommendations, and whether the recommendations had been followed. That gave us the context to begin NASA's planning activities. After we had catalogued and reviewed previous studies, we discussed them with the chairs of committees that had produced each one. Then, we began a process of long-range planning, which evolved into a strategic planning process for all of NASA.

In the report's preface, Ride noted that the space program was at a crossroads. It faced a choice to either focus strictly on the shuttle and station or strike out in a new direction and adopt a visionary goal. Ride's report rejected the former idea: "Without an eye toward the future, we flounder in the present." The report continued, however, that "a single goal is not a panacea—the problems facing the space program must be met head-on, not oversimplified."[31]

The report suggested four initiatives that, if adopted, could restore American leadership in outer space. In the study team's opinion, the options were all of equal importance: "This process was not intended to culminate in the selection of one initiative and the elimination of the other three, but rather to provide four concrete examples which would catalyze and focus the discussion of the goals and objectives of the civilian space program." The following alternatives were discussed:

- **Mission to Planet Earth**—using space assets to study the earth on a global scale.
- **Exploration of the Solar System**—sending robotic science missions to the outer solar system, comets, asteroids, and Mars.
- **Outpost on the Moon**—building a lunar outpost to extend the Project Apollo legacy.
- **Humans to Mars**—sending astronauts to the surface of Mars.

The report noted, "The goal of [the MTPE] initiative is to obtain a comprehensive scientific understanding of the entire Earth System, by describing how its various components function, how they interact, and how they may be expected to evolve on all time scales." Echoing the Bretherton Report, the document suggested that the challenge for MTPE was to develop the capability to predict changes that might occur. The Ride Report endorsed the EOS concept of four polar orbiters—two developed by NASA/NOAA, one developed by the ESA, and one developed by Japan. The study team also recommended the development of five geostationary platforms (three by NASA/NOAA, one by the ESA, and one by Japan). Finally, the report advocated space station–attached payloads to provide unique measurements (e.g., annual tropical rainfall). In its entirety, this integrated system "would measure the full complement of the Earth's characteristics, including: global cloud cover and ice cover; global rainfall and moisture; ocean chlorophyll content and ocean topography; motions and deformations of Earth's tectonic plates; and atmospheric concentration of such gases as carbon dioxide, methane, and ozone."[32]

As the System-Z Concept Study had done previously, the Ride Report's endorsement of MTPE was predicated on the availability of advanced technologies that were not yet available. Acknowledging that Titan IV launch vehicles would be necessary to launch polar orbiters, the document assumed innovations in automation and robotics so that "geostationary platforms [could be] lifted to orbit, assembled at the Space Station, and then lifted to geosynchronous orbit with a space transfer vehicle." The study team determined that the initiative could not be a NASA-only endeavor: "Because of its international and interdisciplinary nature, Mission to Planet Earth requires the strong support and involvement of other US government agencies (particularly the National Science Foundation and the National Oceanic and Atmospheric Administration) and of our international partners. The roles of the various federal agencies have been examined in detail by the Earth System Sciences Committee." The Ride Report concluded that a broad strategy for the American space program was necessary before any decisions could be made regarding the implementation of these initiatives. However, it stated that in the meantime, it was appropriate to conduct preliminary assessments of the initiatives. A careful reading of the report reveals that the two uncrewed science missions received the strongest support—largely because both had been thoroughly examined by the space agency's many advisory bodies and were technologically feasible. As one team member later recalled, Ride was a big earth science fan and "tried to make sure its potential didn't get overshadowed by the higher visibility lunar and Mars initiatives."[33]

A Congressional Budget Office (CBO) study released the following year calculated that fully implementing the Ride Report's recommendations for MTPE would cost anywhere from $20 billion to $25 billion dollars over fifteen years. The reaction around the space policy community to this figure was mixed between positive and pensive. As Assistant Associate Administrator for Space Sciences and Applications Joseph Alexander later recalled, the science community received the report with great enthusiasm: "I think . . . the scientific community liked the fact that several of her options highlighted the value of science to the space program and suggested that a scientific thrust could constitute a raison d'etre for NASA." On the other hand, OMB's Jack Fellows told the *Christian Science Monitor*, "The time is past when NASA [can conduct] a study and a president [will] announce a bold new program." Noting that the space agency was facing difficult fiscal challenges, Fellows indicated that any new initiatives would receive significant OMB scrutiny: "No one should be in a situation where a vision is sold without a clear understanding of its price tag."[34]

## An Announcement of Opportunity for EOS

Administrator Fletcher responded to the Ride Report by creating the Office of Exploration, headed by a new associate administrator to study options for the long-range human exploration of the solar system using the report's evolutionary approach as a guide. In principle, MTPE was not a new initiative. But in reality, the program had stalled in the aftermath of the failure to bring NASA, NOAA, and the NSF together at the first meeting of the CES. The Ride Report induced Administrator Fletcher to open a new dialogue with his counterparts at the other two agencies—NOAA Administrator Clarence Brown and NSF Director Erich Bloch. Through a series of letters, the three leaders agreed that despite earlier difficulties, it would be beneficial to develop a coordinated message on global change research to communicate to the Reagan White House. To that end, on 21 October 1987, they sent a letter to OMB Director James C. Miller III stating their combined support for a federal global change research initiative. The letter stated, "The evidence has been mounting for several years that the Earth's global environment may be changing in ways to which we cannot easily adjust." Essentially reading as a two-page summary of the Bretherton Report, the letter outlined the agencies' joint objectives and suggested that future budget deliberations must give serious attention to the issue. The letter argued, "An effective national program demands the sustained, long-term commitment of all participating federal agencies. . . . We recommend that the FY1989 budget process consider collectively these requests as complementary elements composing a critical national program."[35]

Science Advisor William Graham, to his credit, had not viewed the disastrous first CES meeting as portending permanent failure. He simply believed that a certain amount of bureaucratic wrangling was to be expected and that building some kind of consensus would take time. After the departure of Calio, Graham chose USGS Director Dallas Peck to chair the committee and attempt to bring the community back together. The joint NASA-NOAA-NSF letter confirmed that there was support for a coordinated federal effort in global change research. Peck spent the following months speaking with each of the relevant agencies to verify that they still wanted a committee and determining how they would interact. In December, the committee came together for its first meeting under Peck. USGS Deputy Director Ray Watts (who had been the organization's liaison to the Bretherton Committee) recalled, "The second meeting was just as smooth as could be. They all knew they had been heard, and they all came to the table."[36]

This meeting began a practice that kept the CES busy for the next several months—each representative to the body was given an opportunity to discuss his or her agency's intended role in global change research. The committee also established a working group that would do the bulk of its work, staffed by NASA's Shelby Tilford, NOAA's Eileen Shea and Mike

Hall, and the NSF's Robert Corell. Over the subsequent half year, the working group established a forum for global change research discussions. As NASA's Dale Myers recalled,

> each of the departments briefed the others, just presenting their views of what they were presently doing and what they hoped to do someday. We spread the word, and they spread the word as to what they had to offer. . . . The process served to educate everybody in a way that had never been done before. And as everybody got to know everybody else, the committee took on an importance that [went] far beyond the actual meetings. It [became] a network.

During that period, the NRC's Global Change Committee was integrated into the network. As David Kennedy writes, "The Committee on Global Change (CGC), which had been trying for several years to figure out how to deal effectively with the federal science bureaucracy, also immediately embraced the CES. Before long, CES and CGC members were routinely attending each other's meetings, technically as observers but in fact, very often, as active participants."[37]

At the same time that these efforts to develop effective coordination between the major global change research agencies was ongoing, Administrator Fletcher had received a preliminary briefing on the launch vehicle assessment that he had ordered in July. The outcome of the study was straightforward. Only by using the Titan IV could the current EOS plans go forward without major changes. The vehicle was powerful enough to lift the polar orbiting spacecraft, with sufficient margins for growth in the total spacecraft weight. The Titan III could lift two smaller polar orbiting spacecraft, but this would require eliminating NOAA's operational instruments (and their necessary redundant backups). Using a Delta II would require an entirely new approach to EOS. In addition to the loss of the NOAA instruments, using this smaller launcher would necessitate the removal of radar and laser altimetry instruments and the cancellation of plans for a promising new instrument concept called the High-Resolution Imaging Spectrometer (HIRIS) (a microwave sounder). The impact to the science mission would mean that EOS could not provide data on landmass and vegetation changes, sea surface temperatures, atmospheric chlorine, or polar snow mass. In order to recover these science capabilities, EOS would have to go from a two-satellite configuration to a six-satellite configuration. The study team requested, and Fletcher approved, a more detailed investigation of mixed launch vehicle options in which instrument packages, satellite designs, operating altitudes, and launch vehicle capabilities could be traded so that an optimal configuration for EOS could be developed (or at the very least an optimal alternative to the two Titan IV launches could be examined). These options would be considered for completeness, but the initial assumption was that the baseline Titan IV approach would win because keeping as many instruments as possible collocated on a spacecraft would ensure the simultaneity requirement long envisioned for EOS.[38]

In January 1988, preliminary results on EOS servicing were presented to Lennard Fisk. Mixed teams presented these initial findings, with groups from both the Space Station Program and EOS Project Office offering cost estimates. The EOS team discussed budget options for various spacecraft configurations and the polar platforms they might fly on (even though the latter was not part of their program). The space station team reviewed the required outlays for various polar platform configurations and the spacecraft they might carry (even though the latter was not part of their program). The end results were significantly different appraisals of potential program expenses. One key finding from each group was that the station team believed robotic servicing would be cheaper, while the EOS team concluded that expendable satellites would be less pricey.[39]

Shelby Tilford believed there were some serious problems with the station team's estimates. First, he thought the panel underestimated the development costs and overestimated the capabilities of the needed robotic servicing spacecraft. Second, he was apprehensive about costing that did not include launch expenses for the robotic servicing spacecraft or include any margin for spares. Finally, he was concerned about a servicing approach in which new instruments were added but old instruments were not removed because instruments of lesser quality would be left in space and compromise the platform's operational efficiency. If a more desirable servicing method was selected, whereby new instruments were actually swapped out for the old ones, the servicing costs would be higher because the spacecraft would have to be more sophisticated. When Tilford corrected for these different assumptions, he found that the two estimates were roughly the same. The risk associated with robotic servicing was significantly higher, however, given that it had never been done. As a result of this analysis, a broad consensus was reached within the space agency that the appropriate plan would be to develop two EOS spacecraft that would be launched on a Titan IV and replaced every five years. Over fifteen years, NASA would build and operate a total of six satellites.[40]

In early 1988, assuming that this approach would be selected, the EOS Project Office released the announcement of opportunity for EOS as a means of reaching out to the science community. The desired result would be the identification and selection of researchers to do the actual scientific work. In total, thirty investigation teams would be chosen for instrument-specific and cross-disciplinary investigations (the size of these teams would be determined by the proposers). The previous fall, in advance of the official release, letters had been sent to forty thousand potentially interested individuals. Responses to the formal announcement were due by July 1988. The intention was for the space agency to make its selections by February 1989.[41]

By December 1988, it was clear that NASA's schedule for EOS would push the project's initial launch date beyond what NOAA could accept for its requirements to maintain operational coverage from polar orbit. After several discussions between Lennard Fisk and Thomas Pyke, Fisk sent an official letter to Pyke stating that NOAA's operational instruments would no longer be part of EOS. NOAA would have to procure, in partnership with NASA, the additional spacecraft needed to replace existing satellites. NOAA and NASA would still work together for EOS, but the nature of that arrangement would fundamentally change. As Fisk's letter noted, "NASA will now shift its plans for the Earth Observing System and polar platforms from a research and operational mission to a research and prototype mission."[42]

While these events were transpiring, the Bretherton Committee released its second report. *Earth System Science: A Closer View* was a two-hundred-plus-page opus on the who, when, and why of global change research. This follow-up study expanded upon each of the major topics discussed in the initial review two years earlier. The document essentially duplicated earlier efforts (e.g., the Goody Report, Eddy Report, EOS Science Group Report, EOSDIS Panel Report, and Ride Report) but provided much more detailed analysis. In addition to EOS, the committee promoted earth system explorer missions. These more limited missions would examine very specific climate system elements that were not necessarily a part of the EOS science program. The study team saw them as a parallel to similar exploration projects being pursued around planetary bodies elsewhere in the solar system. These missions were actually already being planned under the direction of Shelby Tilford—who called them earth probes—and would prove to be a key part of the MTPE initiative. Another original contribution of this report was its in-depth programmatic assessment regarding the agencies that were currently involved in research activities that fit into the committee's vision for a federal global change research program.[43]

Around the same time, the NRC's Space Studies Board released a series of reports that had been three years in the making. In 1984, Burt Edelson had asked the panel to "undertake a study to determine the principal scientific issues that the disciplines of space science would face during the period from 1995–2015." This period coincided with the planned operational life of space station *Freedom*. To complete this task, the Space Studies Board had organized itself into task groups covering the following fields: earth sciences, planetary and lunar exploration, solar and space physics, astronomy and astrophysics, fundamental physics and chemistry, and life sciences. While it had planned to complete the study in two years, it had taken an additional year to release its findings. The opening statement of the report that examined earth sciences read, "We now have the technology and the incentive to move boldly forward on a Mission to Planet Earth. We call on the nation to implement an integrated global program using both space-borne and Earth-based instrumentation for fundamental research on the origin, evolution, and nature of our planet, its place in our solar system, and its interactions with living things, including mankind."[44]

The MTPE report outlined a "unified program for studying the Earth, from its deep interior to its fluid envelopes." Observation systems for MTPE were broken down into two categories—satellite-based observations and complementary in situ observations. Stealing a page from the Bretherton committee's playbook, the Space Studies Board supported the development of both next-generation satellites characterized by EOS (including the Ride Report–endorsed constellation of five geostationary satellites) and a series of special missions that were analogous to NASA's earth probes. To pursue this aggressive agenda, the committee called on the federal government to make the following commitments:

- Long-term budgetary support for a vigorous systematic exploration of the earth,
- Maintenance of an open-skies policy,

- International cooperation and coordination regarding all relevant systems,
- Full coordination of all federal agencies at the programmatic and budgetary levels,
- Development of a more comprehensive program in the solid earth sciences within NASA.

Noting previous work done in this area by the board, the report stated,

> The [previous] strategies of the Space Studies Board have emphasized that global Earth science investigations from space will naturally be followed by global exploration for resources and by development of natural-hazard warning systems. Of course, all such applications must be based on thorough scientific understanding. The task group hopes that the scientific return derived from the strategy contained in [previous SSB] reports and in this report will provide a sufficient foundation for all applications.[45]

## Another Battle to Control Global Change Policy

In June 1988, James Hansen appeared before the Senate Committee on the Environment and Natural Resources. In testifying about the effects of greenhouse gases, Hansen said, "The greenhouse effect has been detected and it is changing our climate now. . . . We already reached the point where the greenhouse effect is important." Hansen's testimony came as the American Southwest was experiencing one of the hottest summers in history to that point and the American Midwest was experiencing a prolonged drought. Hansen's testimony linked global warming to these weather patterns, explaining that high temperatures and drought were the first signs of trends that would become commonplace in the 1990s. However, none of the other scientists on the panel would agree that the specific conditions of that summer were the result of global warming trends. Michael Oppenheimer of the Environmental Defense Fund even said, "I'm not yet convinced that doomsday is here." The other panelists mostly agreed that there was sufficient time to abate any serious crisis. They argued that all that was required was to take immediate action to reduce the burning of hydrocarbons—as if it would be easy to slow the ever-growing global demand for these cheap and abundant fossil fuels, which had been the engine of modernity since the beginning of the first industrial revolution.

Fortunately, most of the press coverage highlighted Hansen's testimony. Senator Timothy Wirth (D-CO), who chaired the hearings, was quoted in the *New York Times* as saying, "As I read it, the scientific evidence is compelling: the global climate is changing as the Earth's atmosphere gets warmer. Now, the Congress must begin to consider how we are going to slow or halt that warming trend and how we are going to cope with the changes that may already be inevitable."[46]

The following month, the Senate Committee on Commerce, Science, and Transportation held hearings to discuss whether there was a need to pass legislation directing the White House to create a global change research program. Committee Chairman Fritz Hollings (D-SC) had recently introduced the Global Change Research Act, which would mandate a ten-year federal study. In essence, the legislation sought to create a more substantial and long-term form of the CES. Hollings opened the hearings by saying,

> Hotter weather, severe droughts in the United States and Africa, the Antarctic ozone hole and worldwide thinning of the ozone layer all serve to underscore the need for a national and international strategy to protect the fragile ecology of our planet, this spaceship Earth. . . . This is not a trivial undertaking. The scientific questions are fundamental and difficult because they require us to understand the Earth as a system. We now know these interrelationships are more complex than we first thought. . . . We here in the committee are interested in planet Earth, a "mission to planet Earth," as Sally Ride recommended here not long ago.

Among the witnesses appearing before the committee were Dallas Peck, Francis Bretherton, and Bob Watson. When asked his opinion of the bill, Peck replied, "I have only had a chance to glance at it, Mr. Chairman. I think it's an awfully good start. . . . The goal of the bill, it seems to me, is to institutionalize much of what we are doing on an informal basis in a federal council right now." Peck, Bretherton, and Watson all agreed that a coordinated effort within the federal government was a good idea, but they did not endorse anything more than the CES process currently at work. Peck expanded on his agnostic opinion in his formal written responses to questions from the committee, noting,

> There has not, to the extent that [the CES] can perceive, been money wasted due to a lack of coordination. However, it is hoped that through a coordinated effort the overall effectiveness and productivity of the federal government's programs . . . may be enhanced. [The Global Change Research Act] is consistent with the intentions of the Executive Branch to provide a focus for coordinated planning though [the Federal Coordinating Council for Science, Engineering and Technology].

As a result of the tepid response from the relevant administrative agencies and the scientific community, the Global Change Research Act was not acted upon that year.[47]

In late 1988, the CES working group produced a report called *Our Changing Planet* that was intended to accompany the incoming administration's budget request for the following fiscal year. The report had been requested by Jack Fellows, who wanted to demonstrate how much funding each agency was dedicating to global change research. The agencies had been hesitant to cooperate for fear of exposing themselves to criticism from the current administration. Fellows argued that by contributing to the report, they would obtain an opportunity to present their combined work as something unique and important within the federal government. Paul Dresler recalled,

> Jack said, if you do this, the administration is so eager to come out with some kind of a statement about what we're doing nationally in response to global change, we'll get a statement about the global change program released at the same time as the president's budget. Well, you know, such visibility! We had to meet this challenge. . . . In times past, that's when agencies said, all right, well, we've had enough discussion here. We don't need to share with you how much we're getting here, there, or the other place. Because that's where they're vulnerable; you start laying your program out on the table and suddenly others have an opportunity to see what's wrong with it. It's hard enough having people within your agency taking potshots at programs without opening the door even wider. . . . The agencies wanted to show they were players, so they had to show something, but they certainly didn't want to show it all, because that's when you are vulnerable. You're putting your budget

on the line, and nobody knows where this is going. This is brand new. So most agencies thought, we can risk a certain level; then we'll still survive if for some reason the dagger comes out after it's out on the table.

This is why the agencies contributed only top-level budget figures to the study team. Still, the report showed that the government was spending even less than had been discussed during the recent hearings before Senator Hollings's committee.[48]

In January 1989, the NRC released the report in booklet form. Francis Bretherton later noted that the plan for a U.S. global change research program was so similar to what he and his committee had outlined that it was essentially "our report." The document summed up the why and how of American involvement in global change research and identified three overarching objectives clearly derived from the goal and challenge statements of the Bretherton reports:

- Establish an integrated, comprehensive monitoring program for earth system measurements on a global scale.
- Conduct a program of focused study to improve our understanding of the physical, chemical, and biological processes that influence earth system changes and trends on global and regional scales.
- Develop integrated conceptual earth system models.

The science elements to be studied by a U.S. global change research program would include biogeochemical dynamics, ecological systems and dynamics, climate and hydrologic systems, human interactions, earth system history, solid earth processes, and solar influences. Although NASA would receive only a paltry $36 million combined for both FY1989 and FY1990 under the proposed plan, if it were successfully adopted, the space agency would have resurrected a prominent role in earth system science research from the ashes of the fallen global habitability initiative.[49]

## Bush, the U.S. Global Change Research Program, and MTPE

John Kingdon writes that "a change of administration is probably the most obvious window in the policy stream" because of the far-reaching impacts this change has on political appointees and opportunities for new ideas to grab the attention of leaders during a fast-paced period of change. While space policy was not a hot issue during the 1988 election cycle, Vice

President George H. W. Bush gave it more attention than any presidential candidate since John Kennedy. Bush's interest in the topic stemmed from his meeting with the families of the *Challenger* astronauts several years earlier. June Scobee, the wife of the mission commander, told him at that meeting not to let the disaster hurt NASA and to fight to keep the space program alive. A few months later, at the dedication of a memorial for the fallen astronauts at Arlington National Cemetery, Bush stated that "the greatest tribute we can pay to the *Challenger*'s brave crew and their families is to remain true to their purpose, and to rededicate ourselves to America's leadership in space."[50]

In October 1987, just weeks after declaring his candidacy for the presidency, Bush delivered a major policy address on the future of the space program at the Marshall Space Flight Center in Huntsville, Alabama. He told the crowd gathered at the NASA field center, "In very basic ways, our exploration of space defines us as a people—our willingness to take great risks for great rewards, to challenge the unknown, to reach beyond ourselves, to strive for knowledge and innovation and growth. Our commitment to leadership in space is symbolic of the role we seek in the world." Exhibiting his moderate Republican roots, Bush spoke eloquently of the environmental problems facing humanity and the role that MTPE could play in altering "our disastrous course." Nearly a year later, his campaign released a press statement discussing the candidate's proposals for the space program. Its most important passage read, "I am committed to reestablishing America as the world's leader in space." The document said that President Bush would promote four significant initiatives, and among them was MTPE. Regarding MTPE, the statement read, "The information gained though this project will be of great value to farmers, fishermen, weathermen, scientists, all of us."[51]

Unlike the space program, environmental policy was a central issue during the presidential election campaign. Vice President Bush had successfully linked the two issues by promoting MTPE as essential to altering the nation's disastrous course on the environment. During a campaign stop in Detroit, Bush declared himself an environmentalist and suggested that he would seek to unite the nations of the world in a fight against global warming. Although some in the press believed Bush was trying to steal a Democratic issue, his stand on environmental issues during the campaign effectively distanced him from President Reagan. However, this position was not purely about politics. Bush seemed to genuinely believe that government-led research efforts could be leveraged to remove uncertainties that he believed still existed in climate science. After he defeated Michael Dukakis, the president-elect secured a reputation during the transition period as a leader who was willing to work with the Democratic-controlled Congress to protect the environment.[52]

Almost immediately after taking office, President Bush made the U.S. global change research program a key presidential initiative for the FY1991 budget. By doing so, he single-handedly pushed the program onto the national agenda and provided it with critical presidential backing for the first time. This dramatically increased the pace of activity aimed at gaining the adoption of a major new initiative and left the OMB scrambling for additional information. Paul Dresler recalled, "OMB came back to us and said, well, gee, you guys did a great job. This is fine. Nice work. But we need more." Jack Fellows wanted the CES working group to prioritize the seven science elements discussed in its earlier report (biogeochemical dynamics, ecological systems and dynamics, climate and hydrologic systems, human interactions, earth system history, solid earth processes, and solar influences). Dresler remembered, "For the first time, the Committee on Earth Sciences was going to have to do something akin to designating winners and losers. I have never seen such anguish amongst the committee as I did then." But the program's designation as a key presidential initiative was enough to move the members of the CES past their previous hesitation to provide information. As Dresler explained, "The agencies were saying, 'Presidential Initiative. New money. They're serious—we're going to put some extra effort into this.'"[53]

On 15 January 1989, Senator Hollings reintroduced his legislation to establish a global change research program with the objective of developing predictive capabilities. The following month, the Senate Committee on Science, Commerce, and Technology held a second set of hearings to discuss the legislation. The NSF's Robert Corell testified that the goal of developing modeling tools was already the objective of ongoing global change initiatives within the federal government. He argued that the CES was an effective coordinating committee for those initiatives. The member agencies continued to see the Hollings legislation as simply a means of codifying the committee's charter, and they suggested only minor modifications to the bill. Seven weeks later, the Senate Committee on Commerce, Science, and Transportation held another hearing to examine MTPE more closely. Among the witnesses were Lennard Fisk and Sally Ride. The intent of the hearing was twofold. First, senators wanted to understand the scope of NASA's program. What was NASA's idea for MTPE, and how did it fit with the White House–coordinated global change research program? Second, how much would it cost? The CBO had suggested $15 billion to $20 billion. Was that accurate? And was that too expensive?[54]

The committee and witnesses engaged in a very open dialogue, openly asking each other questions and explaining themselves and frequently going over the allotted time restrictions. Lennard Fisk summed up NASA's position, suggesting that MTPE included EOS, earth probes, attached payloads onboard the space station, and perhaps a future constellation of geostationary satellites. All of these platforms would be used to engage earth scientists in original research, including mathematical modeling of planetary processes and intense data analysis. Fisk did not specify what missions would make up the earth probes aspect of MTPE, as that was to be decided in the upcoming budget process and presented later to the Congress. Likewise, he did not specify any of the anticipated costs for EOS. He did make it clear that the geostationary platforms were not yet part of NASA's planning but were in the earliest conceptual phase. He then summed up:

> [MTPE] is part of the larger US Global Change Research Program, which in turn contributes to an international scientific plan of global research organized by the international Council of Scientific Unions and called the International Geosphere Biosphere Program. The U.S. federal agencies have come together under the auspices of the Committee on Earth Sciences to define the U.S. Global Change Research Program together with roles for each of the agencies.

With respect to cost, it was Ride who led the discussion. Though the cost seemed quite high at first glance, she noted that the CBO estimate covered the whole program over a period extending past 2005—more than fifteen years in the future. Over such a long time, a $20 billion investment by the government was more than justified given what she believed to be the expected returns. Playing devil's advocate, Chairman Hollings wanted to know whether there was more on the other end of this period than a "$20 billion thermometer." Fisk answered by pointing out that the positive impact for future economic policy decisions based on a solid research program would make the investment in MTPE a bargain. In the end, the committee members seemed pleased with this high-level analysis.[55]

On 1 March, Administrator Fletcher sent a letter to President Bush regarding the forthcoming twentieth anniversary of the first human landing on the moon. Fletcher wrote the president that the occasion provided a unique opportunity to define the administration's commitment to the exploration of space. The administrator suggested that Bush's participation in an event planned at the Smithsonian's National Air and Space Museum would enhance the significance of the anniversary. Fletcher suggested, "Taken by itself, an anniversary of this sort tends to focus on past glories and a nostalgia for days long gone. Coupled with a message of leadership and strong direction for the future, it becomes an integral part of the American space experience; it can reenergize the country by setting new challenges and new horizons in the historic context of earlier goals successfully met."[56] This letter persuaded the White House to think of the possibility of utilizing the anniversary, only four months away, as a platform for a major space policy speech. In the meantime, Vice President Dan Quayle, who served as chair of the White House National Space Council, directed administration officials to begin thinking about recommendations for new initiatives.[57]

In early March, Mark Albrecht was selected as the Space Council's Executive Secretary. Albrecht had previously served as national security adviser to Senator Pete Wilson (R-CA) but was a controversial choice because he was not well known within the civilian space policy community. Both Quayle and Albrecht were enthusiastic about setting a new course for the space agency but were concerned about the potential fiscal implications. Given his unease regarding the budgetary constraints, Quayle immediately scheduled a meeting with OMB Director Richard Darman to discuss announcing a new initiative. The meeting took place in early April and was attended by Quayle, Albrecht, Darman, and OMB Executive Associate Director Bob Grady (who had been an environmental policy adviser to Bush during the campaign). The conversation centered on the possibility of announcing an initiative focused on human exploration beyond Earth's orbit, starting with a return to the moon and perhaps continuing on to Mars early in the next century. Darman was supportive but cautioned that the current budget situation mandated that the administration could not serve up an Apollo-like crash program. One important outcome of the meeting was that if such a program were adopted, it would be the central organizing principle for the entire civil space program so that everything from the space station to space transportation to planetary programs would be structured around accomplishing this continuing objective. President Bush agreed with this approach two days later when Quayle presented it to him at their weekly lunch.[58]

On 12 April, President Bush announced that Admiral Richard Truly would be his nominee to replace outgoing NASA Administrator Fletcher. Truly had served as a NASA astronaut for nearly fifteen years and made two flights into space—the second in August 1983 as commander of the space shuttle *Challenger*. After this mission, he had left NASA for several years to head the newly formed Naval Space Command. In 1986, he had returned to NASA as associate administrator for space flight and was widely credited with guiding the shuttle program back to operational status after the 1986 *Challenger* accident. Although he would ultimately run afoul of both Quayle and Albrecht, his nomination was met with widespread approval, and the U.S. Senate easily confirmed him.[59]

It took nearly six weeks for Albrecht to sit down with Truly to discuss the possibility of announcing a new human spaceflight initiative. Truly's initial response was not supportive—he did not think NASA could handle the responsibility so soon after recovering from the *Challenger* accident. He was ultimately convinced, however, that he could not rebuff a presidential request. At the end of May, Truly assembled his leadership team to discuss proposals for the initiative with Albrecht. At this meeting, he expressed his belief that "the real program was Earth, Moon, and Mars as a total program strategy with both man and machines working together. It is that program that I think we need to proceed with." Truly believed MTPE would be a critical part of any new strategic direction for the space agency.[60]

As a team of engineers at JSC set about assembling a program for human exploration of the moon and Mars, Shelby Tilford continued to work hard within the CES on the Earth-based portion. Since naming MTPE a presidential initiative, the Bush White House had been asking for additional budgetary details regarding ongoing and planned projects. This was at the top of the list of activities being carried out by Tilford during this period. The challenge was to provide more detail while at the same time finding ways to protect ongoing missions. The first step in the process was to come up with two overarching categories of research that would characterize the different projects being carried out by each agency. Focused research was defined as "an agency program or activity designed specifically to study global environmental changes or global processes which constitute part of the Earth's environmental system." Contributing research was defined as "an activity which was established and primarily justified on a basis other than the specific study of global change, but which has the potential to contribute substantially to global change research." The CES decided not to explicitly prioritize any research program or scientific investigation, primarily because the committee members could not come to a consensus agreement on absolute rankings. During the first six months of 1989, however, the committee centered its work on separating the various projects into focused and contributing research.[61]

Earlier in the year, as these machinations were ongoing behind the scenes, NASA announced its selections for EOS instruments. For the EOS-A and EOS-B series, the space agency had chosen nearly two dozen instruments with a total of 92 associated instrument-specific or discipline-specific investigations. Additionally, 28 interdisciplinary research teams and more than 500 scientists (representing 32 states, 13 countries, and 168 institutions) were also chosen to pursue research using satellite data from EOS spacecraft. After this announcement was made, NASA's senior leaders ordered a nonadvocate review, which was an independent assessment required before the space agency could officially award contracts for hardware development. During the appraisal process, the EOS Project Office presented all of its project information with regard to program goals, cost estimates, schedules, technical details, program management mechanisms, risks to schedule, and risks to cost to the independent board. By mid-July, this panel had completed its review and reported its recommendations to Lennard Fisk. While the group made over a dozen recommendations, the three most prominent addressed the size of the EOS spacecraft, the number of instruments, and the relationship between the Space Station Program and EOS.[62]

The EOS Project Office had previously decided that each successive platform in the six-satellite series would include instrument upgrades, hoping to build flexibility into the program if the science warranted a change in course. The review team did not support this approach, arguing that it would present data collection and processing challenges that were too difficult to solve. As a result, it recommended reducing the size of the payloads and making all six spacecraft identical. The study group further recommended that only those instruments that had to fly as part of a specific package (to ensure simultaneity with other instruments) should make up EOS. It argued that stand-alone instruments should be removed and either placed on the space station as an attached payload or flown as a separate mission. Ultimately, Lennard Fisk chose to follow these two recommendations—making all six platforms identical and removing nonsimultaneous instruments. These were seen as budget-saving options, as later versions of the satellite were expected to be cheaper to produce. Perhaps most importantly for the long-term management of the program, the outside assessment suggested,

> While one anticipates encountering bureaucratic complexities when dealing with organizations external to NASA, one believes that this should not be the case for internal interfaces. The Space Station Program has evolved to where the EOS program is the predominant, if not the sole, user of the NASA Polar Orbiting Platform (POP). However, the EOS Program has a circuitous and very slow route to influence and/or modify the Polar Orbiting Platform capabilities. . . . The Committee recommends that the POP be decoupled from [Space Station Freedom] and incorporated into the EOS Program . . . in a timely manner.

Because of the inherent inefficiency of the station-EOS marriage, Fisk agreed that they should be divorced. The transition was supported across most of NASA's senior leadership, and the decision was made to transfer management of the polar platform from the Space Station Program to EOS. In September 1989, after these critical decisions were made, NASA presented its EOS budget to the OMB as a new-start proposal for FY1991.[63]

On 20 July, President Bush announced a new strategic vision for the space program before a crowd assembled outside the National Air and Space Museum in Washington, DC. After praising the unparalleled achievement of Project Apollo, he said,

> The time has come to look beyond brief encounters. We must commit ourselves anew to a sustained program of manned exploration of the solar system—and yes—the permanent settlement of space. We must commit ourselves to a future where Americans and citizens of all nations will live and work in space. And today, the U.S. is the richest nation on

Earth—with the most powerful economy in the world. And our goal is nothing less than to establish the United States as the preeminent space faring nation.

In 1961 it took a crisis—the space race—to speed things up. Today we don't have a crisis. We have an opportunity. To seize this opportunity, I'm not proposing a 10-year plan like Apollo. I'm proposing a long-range, continuing commitment. First, for the coming decade, for the 1990s, Space Station Freedom. Our critical next step in all our space endeavors. And next, for the new century, back to the Moon. Back to the future. And this time, back to stay. And then, a journey to tomorrow. A journey to another planet. A manned mission to Mars. . . .

The Space Station will also serve as a stepping stone to the most important planet in the solar system—Planet Earth. . . . Environmental destruction knows no borders. A major national and international initiative is needed to seek new solutions for ozone depletion, and global warming, and acid rain. And this initiative—Mission to Planet Earth—is a critical part of our space program. And it reminds us of what the astronauts remember as their most stirring sight of all. It wasn't the Moon or the stars, as I remember. It was the Earth. [A] tiny, fragile, precious, blue orb, rising above the arid desert of Tranquility Base.

President Bush announced that he was directing Vice President Quayle to "lead the National Space Council in determining specifically what's needed for the next round of exploration: the necessary money, manpower, and materials; the feasibility of international cooperation; and develop realistic time tables—milestones—along the way." Following Bush's speech, Truly held a small press conference at the White House. In addition to discussing the new human exploration initiative, he spoke about MTPE—it was the first time NASA's remote sensing and earth science endeavors had been placed on an equal footing with the human spaceflight program. During the question-and-answer session that followed, however, it became clear that the agency's focus would be on human exploration of the moon and Mars—a program that became popularly known as the Space Exploration Initiative (SEI).[64]

Congressional reaction and press coverage the day after the announcement were not altogether positive, with most of the focus on the lack of budget or schedule specifics for the moon-Mars initiative. President Bush's speech was not received as a Kennedyesque call to action, and as result,

Congress took a wait-and-see attitude. There was cautious optimism about MTPE, which was already a known quantity. Still, supporters of MTPE were concerned that a new human spaceflight program would push the earth science initiative to the side. Despite its efforts to promote the importance of MTPE, the Bush White House was criticized for its handling of climate change. Press accounts painted the administration as divided and conflicted over how to deal with the global warming issue, and Bush was portrayed as a weak leader on the subject. This perception gained credence in Congress during the latter half of 1989.[65]

The same month that President Bush announced SEI, the House of Representatives introduced its own version of a Global Change Research Act. It differed from the Senate bill by directing the federal government to produce "usable information" for policymakers seeking to mitigate the effects of global warming. This would eventually change the debate on the purpose and objectives of a global change research program. Testifying before the House Committee on Science, Space, and Technology, Dallas Peck indicated that his organization was responsible for directing and coordinating a research program, not for establishing policy responses to global change. At a hearing of the Senate Foreign Relations Committee the following month, EPA Administrator William Reilly suggested that there was little to no coordination within the administration regarding how to respond to global change research results. Senators were left with the impression that President Bush was paying little attention to this issue.[66]

In July 1989, the CES released a follow-up to *Our Changing Planet* that laid out specific proposals for the FY1990 budget. For the first time, the study included a comprehensive overview of the amount of money that should be allocated to different federal global change projects. It was a useful complement to President Bush's budget proposal, which was then being hotly debated by Congress. This detailed overview of federal activities was the second in a series of yearly publications under the same title that for several years cataloged the goals of global change research and the contributions being made by different federal agencies. While the first installment had suggested that only $190 million was included in the FY1990 budget for global change research, this updated report showed that the true number was approximately $1.6 billion (including both focused and contributing research). For NASA, $400 million for contributing research projects represented nearly the entire request for the Earth Science and

Applications Division. It included funding for hardware developments for the UARS and TOPEX satellite; operational costs for the ERBE satellite and *Nimbus 7*; Landsat research and data analysis; advanced instrument development; and other atmosphere, ocean, and interdisciplinary research. The only item not included was funds to support NASA field center overhead costs. Internal budget documents concluded, "The NASA Global Change Program consists of the Earth science program in its entirety."[67]

Later that summer, federal agencies began their annual interaction with the OMB to start developing the FY1991 budget. For NASA, this budget request would include a new start for EOS and formal presentation of MTPE to Congress. The EOS request would be used to begin hardware development of the spacecraft, polar platform, and instruments—projects that were already in their preliminary phases after the science and contractor teams had been selected. The pitch to the OMB was fairly straightforward. The agency stated that numerous federal agencies were making commitments to participate in the international geosphere-biosphere program to study and understand global change. The USGCRP had been formulated to meet these obligations. MTPE was NASA's central commitment to this undertaking, and EOS was the central component of MTPE. Thus, everyone was counting on EOS.[68]

NASA's budget request included the development of two satellites, EOS-A and EOS-B. Planned for launch no later than 1997, EOS-A would carry instruments aimed at landmass observations. Planned for launch no later than 1999, EOS-B would carry instruments aimed at atmospheric and oceanographic measurements. NASA had selected a total of twenty-four instruments to fly aboard these two spacecraft. The FY1991 total request of $235 million would be the tip of the iceberg for EOS. The five-year budget horizon identified by NASA suggested that funding would increase to $1.8 billion annually, adding up to $17 billion within a decade and requiring another $30 billion during the fifteen-year life cycle of the satellite series.[69]

As Lennard Fisk had forecast, MTPE had become an initiative that included EOS, earth probes, space station–attached payloads, and space agency research and analysis elements. As such, it did not have a precise budget line of its own. Instead, its various components were intermingled within selected portions of the Earth Science and Applications Division. The foundation for the initiative would be the UARS (planned launch in 1991) and TOPEX (planned launch in 1992). In addition, NASA had plans for three earth probes: the Total Ozone Mapping Spectrometer free flyer mission (planned launch in 1993), a NASA scatterometer instrument to fly on a Japanese research satellite (planned launch in 1993), and the Tropical Rainfall Mapping Mission (planned launch later in the 1990s).[70]

After the submission of the FY1991 budget requests, the CES was asked to examine the various agency proposals and make funding recommendations to the OMB. This was the first budget developed entirely by the new administration, so the group was tasked with ensuring that the budget reflected President Bush's priorities. This was also the first report drafted under the direction of the newly appointed presidential science adviser, D. Allan Bromley, a Yale University physicist and a personal supporter of global environmental research. Bromley saw the CES coordination process as a model for improving the coordination and management of all federal science programs. He later remembered,

> I saw in the Committee on Earth Sciences, personally, what I thought was one hell of a good thing. These were not only disciplines working together to answer a common question, but their projects were overlapping at the right places, and they were eliminating duplication, they were working together as scientists, even as science managers. To actually accomplish something inter-departmental, with answers on the other end, and do it in a really logical way, that's just the way it's supposed to work.

As David Kennedy noted, "Bromley started to build up the [Federal Coordinating Council for Science, Engineering and Technology] structure, [creating] a new collection of . . . committees patterned expressly on the Committee on Earth Sciences." This provides clear evidence of the high regard in which the CES, and its support for EOS, was held early in the Bush administration.[71]

While the USGCRP was bolstered by the heightened role of OSTP and the CES, fiscal responsibility and deficit reduction were President Bush's top priorities. To make sure that the CES recommendations took into account the existing budgetary constraints, Robert Corell would invite committee members to his office to suggest needed increases to key projects and propose areas where cuts could be made. Paul Dresler remembered, "There was quite a good agreement . . . on the general level for agencies. They certainly didn't agree to the dollar, but you [could] see a trend as to which programs [were] being supported and which [were] not."[72]

The biggest winner in this process was NASA. After the review was completed, the agency's total funding level had not changed. What did change, however, was the approach NASA took to categorizing focused research versus contributing research. Earlier in the year, NASA had classified its entire non-EOS earth science program as contributing research. Now it wanted these projects to be considered focused research, which would make MTPE and EOS the largest recipients of focused global change research funding. The space agency also added some new contributing research projects to the mix, mostly technology developments that the space science program was conducting to produce advanced laser ranging sensors and techniques for lunar observations.[73]

On 6 October, however, the OMB came back to NASA with a request for alternative budget ideas. A letter from OMB Associate Director Robert Grady stated, "Because of the extremely uncertain budget environment it is necessary that NASA provide several additional alternatives for EOS in the FY 1991 budget. The level of funding for EOS can have a significant impact on the level and mix of other programs in the [USGCRP]. These options need to be submitted to the OMB by October 17, 1989, to permit the Committee on Earth Sciences enough time to reexamine the [USGCRP] balance." The OMB requested that NASA consider the following options:

**Option A:** Initiate EOS, but on a more modest schedule. Delay the launch and invest significant advance technology development funds to reduce contingency requirements in the instruments, platforms, and data system. Once certain critical risk milestones were reached, the full development of EOS would begin.

**Option B:** Provide funding in FY1991 for only the proposed data system and several critical-pace instruments in response to concerns about the technical readiness of the EOS program and criticisms of NASA's failure to adequately prepare for the use of science data. As in Option A, full development of the EOS would be deferred.

**Option C:** Fund only one platform series. Defer consideration of the second series while exploring the possibility of flying lost payloads on the European and Japanese platforms instead of the current science payloads and require much greater international integration and sharing of payloads, science investigators, and data system support.[74]

Ten days later, Truly sent his response. In addition to a formal letter, he included a collection of charts with the new budget figures and an assessment of the positive and negative aspects of each option. Truly noted the changes that had already taken place as a result of the nonadvocate review, writing, "Please note that the EOS program submitted by NASA in its requested fiscal year 1991 budget has already been scaled back from the original program in order to minimize both the front-end and the run-out costs." As a result, Truly argued,

> Option A does provide a lower fiscal year 1991 cost, but yields a minimum additional slip of half a year in the launch of the first platform. Option B in our judgment has no advantage in that it simply will delay the decision on EOS while providing no significant information. . . . Option C is the most difficult to analyze since it depends on our international partners flying even more of our instruments than are currently planned. Even if such additional commitments can be obtained, Option C would result in the loss of certain key observations which are essential to understanding global change. In summary, I feel that the EOS program request, as originally submitted, provides the best balance of funding and achievement consistent with the President's emphasis on understanding the Earth's global environment. I do not consider that Option B and C are viable alternatives for fulfilling our nation's commitments to understanding global change. If NASA's original submission is not achievable, Option A, or some variation of it, may deserve further consideration. I am very pleased by the constructive process by which we are determining how best to implement EOS . . . and I look forward to discussing this matter with you personally.

It was clear that Truly believed the space agency was negotiating from a position of strength; thus, it could defend its original plan as both appropriate and necessary given the amount of attention being paid to global change and the environment. He wrote in his letter to Grady, "As you know, there is a strong public, political, and international expectation that the President will move aggressively to initiate the Mission to Planet Earth program in order to understand global change and have the basis for sound policy decisions. I am concerned that if we delay . . . we will call into question the President's commitment and subject other space initiatives to criticism." This behavior was a textbook case of an agency rushing to take advantage of the opening of a policy window. Grady later recalled how President Bush reacted to NASA's proposal for EOS: "I think President Bush had a very science-based orientation on this matter. . . . And . . . his view was that whatever happens should be dictated by the science, and we should keep trying to learn more. So he [favored] a fairly large, integrated Earth science program."[75]

On 27 October, Grady and Truly met face-to-face to discuss NASA's submission. They agreed that the agency would provide additional variations of each alternative so that the CES could have more details for its consideration. NASA responded four days later, providing the OMB and CES with two slight variations for each option. As a result, the administration had three families of options to choose from. Truly's savvy bureaucratic wrangling successfully took the second and third options off the table, which meant the CES essentially had to select the first option because it represented the least amount of change to EOS. With this decision made, NASA finally had presidential approval for EOS. After years of struggle, the USGCRP and MTPE were firmly established as national priorities, and the civilian space program would play a central role. As one commentator wrote, this was "an important turning point for the geophysical sciences with immense consequences for their development. The history of this paradigmatic shift is lost in the dimming memories of the participants in innumerable meetings and debates. It is clear, however, that an overriding factor [in the adoption of the initiative] was the feeling that society could not tolerate 'business as usual.'" At the time, EOS was planned to include three pairs of polar-orbiting satellites, each equivalent in size to the Hubble Space Telescope, along with a complex data archiving system that was intended to enable real-time access to EOS data by scientists around the globe. The program was expected to cost $17 billion during its first decade and as much as $50 billion by 2020. If successfully implemented, this would make it one of the greatest Big Science programs in American history.[76]

## CHAPTER FOUR

# Rearguard Action to Save EOS

In 1990, after years of political infighting, NASA finally gained presidential approval to develop a satellite system to study global change and assess its impact on human society. The problem, policy, and political streams were joined, and MTPE/EOS was adopted. What the agency did not know was that it would face many more years of additional struggle before any operational platforms would actually begin returning data to researchers. NASA was about to run a gauntlet with numerous external review panels and the U.S. Congress gunning to radically reduce the functionality of EOS. At the same time, the space agency came under increased scrutiny and criticism after a series of dramatic technical setbacks in both its science and human spaceflight programs. As NASA's leaders moved forward, they would need to exercise all of their political and bureaucratic acumen to ensure the survival of MTPE and EOS.

### The Year That Changed NASA

Although Richard Truly had successfully gained approval for including EOS in the FY1991 budget, there were serious concerns within the Bush White House regarding the administrator's political competence in guiding SEI. While NASA, the OMB, and the CES were coming together on EOS, the relationship between the space agency and the National Space Council was in the process of complete disintegration. In July 1989, NASA had decided to conduct a study to determine how best to implement SEI. Mark Albrecht had met with JSC Director Aaron Cohen, who would lead this effort, and demanded that the space agency provide the administration with multiple programmatic alternatives. Albrecht told Cohen, "There has to be more than one way to do this. Give us a Cadillac option, then give us the El Cheapo alternative, with the incumbent risks. Talk about all the different technologies that could be learned." Despite these instructions, NASA ultimately presented the administration with a single program architecture that would have cost more than $500 billion to implement. Mark Albrecht later said that he and Vice President Quayle "were just stunned, felt completely betrayed. . . . The [study] was the biggest 'F' flunk, you could ever get in government." Similar to what had resulted from the NASA-OMB-CES negotiations regarding EOS, the space agency had failed to provide President Bush with any real alternatives. Once the massive price tag for SEI leaked, the initiative went into a tailspin from which it would never recover. Perhaps more importantly, the National Space Council used this incident to effectively take over the policymaking apparatus within the civilian space program. This entrenchment of the already hostile relationship between NASA and the White House dramatically reduced Truly's ability to effectively lobby for space agency priorities.[1]

The setbacks in the human spaceflight program complicated the path forward for EOS. After consulting with his National Space Council colleagues, Allan Bromley decided that the NRC should review the entire USGCRP budget. On 29 January 1990, Bromley wrote a letter to Frank Press requesting that his organization answer a number of questions related to the implementation of EOS:

- Will EOS collect the environmental data required for the USGCRP?
- How important is data simultaneity to the ultimate utility of the data collected by EOS, and can the requirement of simultaneity be applied more narrowly than proposed?
- Are EOS platforms, as currently configured, the optimal means for collecting the required data, or are there better alternatives that are more cost effective or timely?
- Does the proposed EOSDIS represent the appropriate approach to support the long-term data collection and modeling effort?

Little did NASA know at the time, but this would be just the first in a long line of challenges to the program that the agency would face in the coming years.[2]

In April, the NRC released its review of the USGCRP and EOS. The review was conducted under the auspices of the Committee on Global Change, the permanent committee formed after the publication of the earlier Eddy Report, which was still chaired by John Eddy. The report concluded that the FY1991 budget plans for the USGCRP defined "an appropriate first step toward a sound national program to reduce the scientific uncertainties associated with global change issues."[3] As requested, a specific portion of the report was dedicated to a review of EOS. A panel chaired by D. James Baker of the Joint Oceanographic Institutes conducted this evaluation. Baker was an experimental physicist who had previously held teaching positions at Harvard University and the University of Washington and for nearly a decade had led NASA efforts to develop oceanographic research projects that would use space-based assets. His subcommittee opened its report with general support for EOS: "It is the view of the panel that the initial investments, though heavy, are both prudent and unavoidable, given [that] continuous, long-term, space-based observations of fundamental environmental parameters are essential to achieving the underlying goals of the program." The space agency's specific plans for EOS-A and EOS-B, however, were challenged. In particular, the group questioned the value of simultaneity:

> Scientific arguments for simultaneous measurements have been developed by NASA for two specific research areas: the role of clouds in climate and the fluxes of the trace gases. With regard to these two cases, we conclude that the number of instruments that must fly together requires at least one large satellite. . . . Measurements to carry out the [USGCRP] emphasis on the role of clouds and the fluxes of trace gases are planned for a series of large spacecraft called the EOS-A series. Scientific arguments for simultaneity in terms of the research objectives of the second proposed, large EOS-B satellite have not been developed, and it appears that these objectives could be achieved with a number of smaller, independent satellites. In principle, the science investigations proposed for EOS-B could be done by a suite of smaller satellites. Since a number of the instruments do not require extensive development, these could perhaps be launched sooner.

While the panel accepted the plans for the EOS-A platform, it strongly questioned the necessity of the EOS-B platform. As a result, Baker's team questioned NASA's approach to identifying launch vehicles that would place earth science satellites into orbit. The group wrote that while NASA's engineers were focused on large spacecraft, "we believe it would be prudent at this time to continue to consider a mixed launch vehicle scenario so that the scientific return of instruments currently designated for EOS-B can be increased or achieved sooner. The option should not be eliminated solely on the basis of consideration of launch vehicles."[4]

The Baker panel clearly favored smaller spacecraft modeled on the earth probe missions endorsed in an earlier Space Studies Board report. The group wrote that these smaller satellites "have considerable potential for providing high priority precursor measurements to EOS. They can advance the time in which some of the measurements critical to understanding global change could be made." The panel also suggested that smaller platforms might be a better bet if budget constraints were to arise: "It would be preferable to delay the launch of EOS rather than to forego or diminish the effectiveness of these near-term projects." In an interview with the *Wall Street Journal* conducted after the report was released, Baker said, "We can't wait for EOS to get those measurements."[5]

This opposition to NASA's EOS approach was part of a growing movement within the scientific community, with others also calling for smaller satellite missions. As just one example, James Van Allen, a giant in the field, said, "The grandiose scale disturbs me. They're creating a monster." There was growing concern that the complexity of getting ten to sixteen instruments to work together on a single EOS spacecraft all but guaranteed cost overruns and schedule delays. This was troubling because both EOS platforms already had long development schedules, which could put the entire mission at risk. Numerous earth scientists feared that EOS would arrive too late to measure the impact of temperature changes during the 1990s, which many believed was crucial to quickly determining whether observed climate changes were the result of human activity.[6]

Don Anderson, who had chaired the 1988 Space Studies Board MTPE study, articulated the concerns of scientists who were troubled that the USGCRP was not properly balanced. With so much funding going to NASA to provide space-based measurements, he and others worried that there would not be enough support for ground-based and in situ (e.g., aircraft, balloon, and buoy) measurements. This possibility was disturbing because the ground-based, space-based, and in situ data were necessary complements for each other. As a result, Anderson said, "we're going to spend ten times as much money and probably do it one-tenth as well as we could, and we won't get anywhere near the results because a lot of things are being left out." Finally, others within the scientific community were uneasy about clustering so many instruments on one satellite that could be destroyed if its Titan IV launch vehicle suffered a catastrophic failure. As Carl Wunsch complained, "It's the shuttle all over again, all our eggs in one basket." This rebellion within the scientific community was extremely worrisome, particularly because the space agency was already facing scrutiny in other mission areas.[7]

Over the following weeks, a great deal of focus within the space policy community was placed on SEI. Despite the poor reception that the initiative had received on Capitol Hill, the Bush administration was still trying to save the program and push through a 24 percent budget increase for the space agency, which represented the strongest action by an American president for a space project during the post-Apollo era. Unfortunately, Senator Barbara Mikulski and Senator Jake Garn (chair and ranking member of the key appropriations subcommittee) were skeptical and specifically warned the White House that SEI was vulnerable in the existing budgetary environment. Regardless, Bush continued to actively promote the initiative—both publicly and with key congressional power brokers.[8]

On 1 May, the Bush administration convened a space summit attended by sixteen members of Congress. President Bush opened the meeting by affirming his personal commitment to the American space program, which he believed to be of vital importance to the nation's future. He appealed for congressional support for his requested increase in spending for the civilian space program, which he asserted would put the program on the path to recovery after many years of underinvestment. He argued that both SEI and MTPE embodied NASA's core mission. Vice President Quayle then launched into a passionate defense of SEI, stating that it was in the national interest to implement a new round of human space exploration.[9]

While most of the attendees were committed to the space agency in general terms, not one member of the congressional delegation stated support for SEI. Representative Robert Traxler (D-MI) and Senator Mikulski both made it clear that they were committed to addressing space station and EOS funding issues before allocating any new funding for human spaceflight. One senior congressional aide recalled, "By this time, Chairman Traxler was carrying the message around that 'we can't afford this given our allocation. We can't do it.' He was already negative about it; so coming out of there I don't think he was convinced differently. Congress had already pretty much made up its mind [about SEI]."[10]

Congressional support for MTPE was significantly stronger, although support for EOS was not as robust as NASA leadership had hoped. During congressional hearings convened to consider the space agency's EOS budget request, Traxler said, "Mission to Planet Earth in my judgment . . . is an immensely popular program. I will tell you. It really has people's imaginations fired up . . . even though it's a very expensive program." On the Senate side, however, Mikulski was raising questions about the size of the EOS platforms, the risk of having so many instruments aboard one satellite, and the potential benefits of placing instruments on smaller platforms. While MTPE seemed to have broad-based backing within Congress, NASA's preference for large EOS satellites was now being called into question.[11]

By early summer, nearly all of the good feelings about MTPE had evaporated as NASA was shaken by back-to-back technical and publicity disasters. On 26 June, the space agency held a press conference to reveal that its engineers had discovered a crippling flaw in the Hubble Space Telescope's primary mirror. The defect meant the largest and most complex civilian orbiting observatory ever launched would not be able to view the depths of space until a permanent correction could be made, which would likely have to wait two to three years for a shuttle visit with newly manufactured parts. Although many of the instruments aboard Hubble would still be functional, the wide-field and planetary camera would be inoperable (reducing by 40 percent the planned scientific work of the platform). Project managers announced that they suspected the problem was in one of two precisely ground mirrors, although they were not sure which one. The two mirrors had tested perfectly on Earth, but once in orbit, they had failed to perform together as expected—they had not been tested together on the ground because of the huge potential expense and the inability to replicate a zero-gravity environment. Lennard Fisk disclosed that the agency was forming a review board to investigate the problem.[12]

Two days after the Hubble revelation, NASA was forced to ground the entire space shuttle fleet after the *Columbia* and *Atlantis* orbiters developed mysterious and highly dangerous hydrogen leaks. Combined with the Hubble announcement, this effectively killed any momentum generated by recent administration activities designed to garner support for space agency programs—specifically SEI. Press coverage was typified by a *Washington Post* article, which argued that the failure of the telescope, which two months ago rode into space amid great fanfare in the hold of a Space Shuttle, led more than a few Americans to wonder whether their country can get anything right anymore. The questioning became even more poignant . . . when the National Aeronautics and Space Administration announced that the shuttles, too, would be grounded indefinitely because of vexing and dangerous fuel leakages. [These problems] may foster beliefs that the United States is a sunset power, incapable of repeating its technological feats of the past.[13]

By early July, the view from within the Bush White House was that the space program had been terribly crippled. In his memoirs, Quayle wrote, "The Shuttle seemed to be grounded all the time with fuel leaks; the mirror on the Hubble telescope couldn't focus; and the agency was pushing a space station design that was so overblown it looked as if we were asking to launch a big white elephant. The mood at the Space Council was grim. . . . I was searching for a solution for NASA." On 11 July, Quayle invited a group of space experts aboard Air Force Two for a meeting to discuss the systemic problems with the space agency. He asked for opinions regarding what action, if any, the administration should take. During the meeting, an idea emerged to establish a task force to examine how the space program could be restructured to better support an era of sustained long-term space operations.[14]

Six days later, Quayle hosted a second meeting at the White House with senior administration officials Bill Kristol, Mark Albrecht, Richard Truly, and White House Chief of Staff John Sununu to discuss the procedures to create such a panel. Quayle recalled,

I wanted [the study] to get NASA moving again. If that was going to happen, then the commission had to have the authority to look into every aspect of the space agency. The result was a long negotiation about the commission's scope. Truly's original position was that it should look only at the future management structure of NASA—that is, what would come after the space station was built. "No," I said, "it will look at the current management situation." Truly next tried to exempt programs from review, and I said "No, programs will be reviewed as well." He asked that the space station be "off the table" and said that we would all be better served if both it and the Moon-Mars missions were off limits to the commission. In other words, the commission shouldn't pay attention to all the most important things we were trying to do during the next couple of decades. "I'm sorry," I said to Truly, "but everything is on the table, and let the chips fall where they may. . . ." I tried to soothe Truly's feelings by making the commission report through him to me.

At the end of the meeting, Quayle had directed Truly to put together an outside task force to consider the future long-term direction of the space program. A little more than a week later, the Bush White House announced that Martin Marietta Chief Executive Officer Norm Augustine had been selected to chair the Advisory Committee on the Future of the U.S. Space Program. The twelve-member task force was charged with reporting its findings within 120 days.[15]

During the 1988 presidential campaign, George Bush had pledged that he would not raise taxes to shore up the support he needed from economic conservatives within his own party. Two years later, however, this pledge was in jeopardy as a slowing economy, growing budget deficits, and a $124 billion bailout for the savings and loan industry had placed the administration in a precarious position. Congressional Democrats believed that Bush's FY1991 budget request had gone too far in cutting spending and not far enough in raising revenue for the government. On 26 June, the White House agreed to a tax increase if steps were taken to reduce federal spending. President Bush released a statement arguing, "It is clear to me that both the size of the deficit problem and the need for a package that can be enacted require all of the following: entitlement and mandatory program reform, tax revenue increases, growth incentives, discretionary spending reductions, orderly reductions in defense expenditures, and budget process reform." Although spun in bipartisan terms, this incredible act of political courage was very likely the proximate cause of his failure to win reelection two years later.[16]

By October, the Bush White House and congressional Democrats had finally concluded their budget negotiations. For the space program, the big loser was SEI—the entire program was eliminated from the budget. Congressional staffer Stephen Kohashi later said, "The primary concern regarding the FY1991 NASA appropriation was dealing with the rising cost profile of previously initiated projects such as the Space Station, Earth Observation System, and space science missions. No one had the stomach to commit to another program start, no matter how modest the initial price given the relative magnitude of out-year costs." MTPE was much luckier. The original request was $661.5 million, and this amount was cut by only $119 million. On 16 November, the authorization committees in both congressional chambers passed legislation granting the agency permission to utilize $132 million (which matched NASA's initial request) to develop the EOS platform, and $35 million was also allocated to develop earth probes. NASA was directed to offset the overall MTPE budget cut by reducing the amount spent on other projects. Thus, within eighteen months, President Bush's two major space initiatives had met with opposite fates—SEI had faded away, but MTPE seemed alive and well.[17]

On 10 December, Norm Augustine presented the findings and recommendations of his advisory panel to the full Space Council. The Augustine Report's most notable finding was that there was "a lack of a national consensus as to what should be the goals of the civil space program and how they should in fact be accomplished." The study team made the fashioning of such an agenda its primary task. The panel argued that the space agency's goals should center on improving global economic competitiveness and environmental security—rationales such as national prestige and national security would be of secondary importance for the first time since the dawning of the space age. If adopted, this emphasis shift would be of momentous importance. In making this recommendation, the panel suggested that it did so with fiscal constraints in mind: "The question thus becomes one of what can and should the U.S. afford for its civil space endeavors in a time of unarguably great demands right here on Earth." In answering this question, the report noted,

> Presumably reflecting public support, both the Executive Branch and the Congress have recently shown a willingness to increase civil space spending on the order of 10 percent per year . . . for a well executed program. This, therefore, is the baseline selected by this Committee to assure at least a first order fiscal test of our proposals. . . . Our specific assumption is that the civil space program will grow by *approximately* 10 percent per year in real dollars throughout most of this decade. . . . This is a budget that can enable a strong space program.

What is shocking in retrospect is that these funding levels would have increased the NASA budget to approximately $30 billion annually by 2000.

Given the economic and fiscal realities of the day, it is puzzling that the panel did not question whether this budget was politically or fiscally realistic. The proposed increase was intended not solely to enable new missions but also to solidify the foundation of existing plans: "NASA is currently overcommitted in terms of program obligations relative to resources available—in short it is trying to do too much, and allowing too little margin for the unexpected. As a result there is a frequent need to revamp major programs." The committee recommended a redesign of space station *Freedom* to focus on life science research and a sooner rather than later phasing out of the Space Shuttle Program (and the replacement of the shuttle with a better vehicle).[18]

While the panel supported a balanced program, it suggested that the space agency's science mission should be the primary focus of future efforts: "It is our belief that the space science program warrants highest priority for funding. It, in our judgment, ranks above space stations, aerospace planes, manned missions to the planets, and many other major pursuits which often receive greater visibility." Space science was not limited to astronomy or planetary science but included the studies necessary to improve fundamental knowledge of earth systems: "Having thus established the science activity as the fulcrum of the entire civil space effort, we would then recommend the 'mission oriented' portion of the program be designed to support two major undertakings: a Mission *to* Planet Earth and a Mission *from* Planet Earth. Both, we believe are of considerable importance." The Ride Report three years earlier had suggested that scientific research could be a justification for NASA that would stand on an equal footing with human spaceflight. The Augustine Report went a step further, making science the primary justification for NASA: "[We] share the view of [President Bush] that the long-term magnet for the manned space program is the planet Mars—the human exploration of Mars, to be specific. . . . [But it should be] tailored to respond to the availability of funding, rather than adhering to a rigid schedule."[19]

"[MTPE] promises a major step in the development of the science and technology of global remote sensing of our planet. . . . The enormous benefits of this information to society require that NASA ensure that the program is well designed and efficiently managed." Thus, while the panel strongly supported the initiative, it did not strongly support the space agency's approach to implementing the program:

> The Committee . . . concludes that the design of EOS must involve a variety of different spacecraft to meet so complex a set of requirements. . . . Particular diligence will be required to assure that the complexity of EOS is controlled. [The Committee recommends] that the multi-decade set of projects known as Mission to Planet Earth be conducted as a continually evolving program rather than as a mission whose design is frozen in time. A combination of different size spacecraft appears to be most appropriate to meet the needs of simultaneity, accuracy, continuity and robustness. The Committee believes that a review of the decision-making process for Mission to Planet Earth, including its relation to the U.S. Global Change Research Program, should be carried out for the National Space Council by a group from government, industry, and academia, headed by the Director of the Office of Science and Technology Policy (OSTP). The review should consider interagency aspects, the role of the [Committee on Earth and Environmental Sciences], international dimensions, and make recommendations aimed at ensuring the success and continuity of the program.

This recommendation was a crushing blow for the NASA planners dedicated to EOS—and the situation worsened. As discussed above, the panel set as its baseline an annual budget for NASA of $30 billion. The panel concluded that if the space agency could not obtain authorization from the administration and Congress at this level, then MTPE and SEI should be scaled back (if not eliminated). By this time, SEI was already rapidly fading away, which meant that MTPE would be number one on the chopping block.[20]

While reaction to the Augustine Report was muted, it was largely supportive. Mark Albrecht thought the findings were "a little milder than we had hoped for and anticipated, but were considered quite strong in the community at large." Dick Malow, the most important congressional staffer for space issues, recalled later that it was "generally considered to be an excellent study. It was well received, particularly because there was a strong emphasis on science." Representative Bill Nelson (D-FL), chairman of the House Space and Aeronautics Subcommittee, called it "the report of the decade." NASA's internal reaction was unsurprisingly negative, as the report painted the space agency as a stagnant organization with bad management and little inspiration. While praising the study in general terms, Admiral Truly expressed his reluctance to condemn the shuttle. Instead, he urged the Bush administration to consider building a fifth orbiter and to preserve the capability to build additional spacecraft. This signaled a continued disconnect between the strategic directions favored by the Space Council and NASA. The commission's recommendation that the Space Council take the lead in independently assessing EOS served only to make the relationship between NASA and the White House even more tense.[21]

On 16 November, after two years of debate, President Bush signed the Global Change Research Act of 1990. Among other things, this legislation transformed the CES into the Committee on Earth and Environmental Sciences. Despite this new congressional mandate, the strong interagency process that had been such a source of strength for NASA and EOS was showing signs of collapse. The agreement that the OMB had struck with the participating agencies was that the funding requests made by each for USGCRP-related activities would be essentially fenced off from the rest of that agency's budget. As David Kennedy writes,

> Agencies either got their money for the programs they'd agreed to or they didn't get it at all: they could not, as they could under the ordinary budget process, petition OMB to reprogram the funds for other purposes. Once all the papers were signed between the agencies and the [Committee on Earth and Environmental Sciences], OMB's rules made Dallas Peck [of the USGS] and the [Committee on Earth and Environmental Sciences]—at least for the purposes of the global change budget—the equal of the cabinet secretaries.

Secretary of the Interior Manuel Lujan disliked this arrangement because it made Peck his equal in budget negotiations despite the fact that the USGS was a DOI bureau. That fall, Lujan had decided to reduce the size of his department's global change budget and apply those funds to other projects. The OMB had informed him, however, that this was not part of the deal. Lujan, by all accounts, "went off like a rocket." Allan Bromley was approached by an angry Lujan asking why "the peons at [the Committee on Earth and Environmental Sciences] were telling him what he could and couldn't do with his budget." Bromley had no interest in getting into a fight with a cabinet secretary and actually would have liked to eliminate the committee because it reduced his authority, but his hands were tied by legislative language that affirmed the primacy of the committee in defining global change research budget requirements.[22]

That same month, a budget agreement reached by President Bush and Congress served to amplify the existing tensions. The deal provided the administration with an opportunity to reexamine existing agency plans and suggest nontrivial changes before submitting its FY1992 budget in February 1991. The OMB immediately began this review and suggested a range of possible reductions for all participating agencies, although each agency would receive an increase compared with the previous fiscal year. Regardless, this review led to an internal rebellion, with most of the agencies refusing to prepare budget submissions based on the most severe budget cut options. The OMB leadership was furious. The director of the Science and Space Programs Branch, Jack Fellows, later recalled, "It made a lot of people here extremely angry. We had a contract with them to deliver on those bottom two options, and all we got was a one-page thing that said, you don't have a program [at that level.] That was outrageous. And there was no way that we could deliver the kind of money that had been talked about [before]." The agencies represented on the Committee on Earth and Environmental Sciences had a closed-door meeting to discuss how to deal with their new budget allocations. As one participant involved in the process later recalled, "Even players like the Department of Agriculture said this isn't the program we've been promising the community. We've got to appeal it." Before any appeal could be filed, however, Allan Bromley got wind of these efforts. He called Dallas Peck and the NSF's Robert Corell into his office. Peck tried to convince Bromley that the OMB had broken a budget deal made the year before. He argued that the science adviser should stand up to the budget office. "Bromley was not interested. Bromley was furious. 'Don't you dare appeal,' he said. 'Appeal and we'll shut the program down.'" This message was carried back to the Committee on Earth and Environmental Sciences, and the matter was closed. However, the tensions remained.[23]

During budget negotiations the previous year, NASA had been able to leverage interagency support to ensure that EOS cuts were off the table. A year later, this support was no longer assured. Jack Fellows recalled,

> Between FY1991 and FY1992 the world changed. Bob Corell and Shelby Tilford and Mike Hall, they [didn't] know that world. They [didn't] get

involved in the politics; the budget agreement. They [didn't] know how all this stuff ties to the bottom line. That's not a criticism, that's just reality. EOS, for the first time, was on the table. The budget pressures finally forced the working group to focus on EOS.

It didn't take long for NASA to feel the pain. Although during these new negotiations, the USGCRP budget earned an increase of $300 million (for a total of $2.4 billion), the EOS budget was cut by $180 million for FY1992. Furthermore, NASA would be forced to reduce the total budget for EOS from $17 billion to $15 billion through 2000. The more assertive leadership of the Space Council, combined with the recommendations of the Augustine committee, was clear in the OMB's guidance to NASA. The organization directed "an external engineering review of the EOS platform configuration and launch sequence, [to be undertaken by] a group of government, industry, and academic experts."[24]

These initial changes did not affect the size of the EOS spacecraft or the number of instruments planned for the six platforms. Instead, they forced the agency to plan on schedule slippages that would move the launch of EOS-A from June to December 1998. By this time, the instrument selection for the satellites was nearly complete. The final budget had been a boon for those advocating an increased budget for earth probes, as this program received an additional $270 million over five years. These funding increases would support the Tropical Rainfall Measuring Mission and Total Ozone Mapping Spectrometer missions, which were already in the planning stages. In the meantime, NASA was able to keep both the UARS and TOPEX projects on track for their scheduled launches in 1991 and 1992.

## Continuing Struggles to Save MTPE/EOS

On 20 March 1991, Norm Augustine appeared before the House appropriations subcommittee that oversaw NASA to describe the findings of his panel. Although three months earlier, he had reached an informal agreement with Dan Quayle to avoid rank-ordering potential missions, as Bob Traxler (chair) and Bill Green (ranking Republican) questioned him, it quickly became apparent that the panel had clear preferences:

Traxler: If the committee cannot deliver on 10% real growth . . . can we assume from your report that the space science strategic plan retains a higher funding priority than the new lift launch vehicle, perhaps?

Augustine: The committee would clearly give highest priority to the space science program, above new launch vehicles or above most anything else one might have. The new launch vehicle . . . would be the next highest priority on our list of activities.

. . .

Green: Ten percent real growth per annum in the NASA budget does not look like a likely prospect over the next few years, given the budget submit that was enacted into law last October. . . . Where does that take us and what would you then recommend?

Augustine: . . . The important aspect is that we not ask NASA to take on tasks if we haven't provided them with enough resources. So, we should actually cut the program.

Green: If we are in a budget crunch, and we do have to have some contraction of the Mission to Planet Earth program, do I infer from your recommendations, basically, that you are in agreement with the National Research Council report for some early probes as soon as possible, followed by the big platform, but that we not have the second big platform, but break it up into smaller free flyers? Is that accurately described?

Augustine: I think that is reasonably accurate.

Green: And if we do run into some budget constraints we should go ahead with the early missions such as UARS and other probes, and achieve savings by putting off the big platform?

Augustine: I think that would be a reasonable decision.

This brief dialogue was the extent of the committee's discussion of MTPE and EOS. Augustine focused mainly on his panel's recommendations for human spaceflight, including moving beyond the shuttle and redesigning the station. On 9 April, the subcommittee held an additional hearing to question senior NASA officials. Traxler and Truly engaged in some verbal sparring over the meaning of the Augustine Report's recommendation for maintaining science as NASA's highest priority, with Truly repeatedly telling the committee, "When Augustine says to fund science at approximately the current fraction, I believe that should be done in a reduced budget environment as well as a growing one." If NASA's budget was going to be reduced, the former shuttle astronaut did not want science to be treated as sacrosanct.[25]

The focus of the subcommittee in this second hearing was clearly to determine the magnitude of any FY1992 budget increase for the space agency. The members were concerned about the practicality of a large increase given the existing fiscal environment. They were particularly interested in determining what should happen to the shuttle and station programs in light of the Augustine panel's recommendations. Traxler methodically went through the entire budget request, allowing NASA's senior leadership to explain the progress of ongoing projects and the expected impact of new research. With the exception of the station, the committee only scratched the surface of most projects. When the hearing turned to MTPE, the members were interested in schedules and the potential benefit to global studies of mixing earth probe missions with EOS. Traxler expressed his consternation about the ongoing controversy regarding the proper balance to Lennard Fisk:

> Traxler: Over the past year or so we have continued to hear a kind of ground noise controversy bubbling about the program, which is still ongoing, I guess. Last year the National Academy reviewed the program, suggested the EOS-A platforms should proceed as designed. In other words a large group of instruments that receive simultaneous data, but the EOS-B platform should be reviewed for possible breaking up. [NASA] suggests that an external engineering review will be undertaken during 1991 to examine additional alternatives to flying EOS instruments. What is the situation now on this issue and why does the controversy continue, and will the external engineering review in any way affect the EOS-A platform?
>
> Fisk: I don't believe it will. . . . The primary thrust of this external engineering review is to look at follow-ons to the first EOS-A platform. There is a lively debate as to how best to proceed in this regard, particularly with the EOS-B series.
>
> Traxler: What is the basis for the ongoing, for want of a better term, controversy, relative to A versus B? What is happening here? Why can't we get this kind of settled? What are the things that seem to create this dissent?
>
> Fisk: I think there is a sense that if you weren't driven by these very valid science requirements for the large platform that flying multiple spacecraft would have an inherent robustness to it that appeals to some people. Some people would prefer that when we go to EOS-B we not go on the large platform but go to the small. But we keep reminding

> people that there is no launch vehicle on the west coast that is capable of launching small or intermediate size platforms. . . . If this engineering panel tells us something that no one has thought about we certainly will welcome that advice. But the criteria would have to be that they present us with something which has a significant advantage in either schedule, cost, science return, or risk reduction. I think the probability is that no one will have advice that will satisfy those criteria.

In the end, the various hearings did little to forecast whether EOS would face significant changes given the existing fiscal constraints.[26]

On 8 May, it was the Senate's turn. Barbara Mikulski convened her Senate Appropriations subcommittee and made it clear during her opening statement that NASA was facing a difficult situation:

> Under last fall's budget agreement, all domestic discretionary spending will increase by only $10 billion. In the President's budget request, what NASA could receive is $1 in every $5 for the program increases. We think that is going to be a little hard to do. I prefer to give NASA its full request, but we know we will be in a very tough budget situation. . . . The most serious issue NASA faces in this climate of tight budgets is the future of the space station.

The committee's question-and-answer session with Administrator Truly focused almost exclusively on the station but was cut short for a welcome reception for General Norman Schwarzkopf. No questions were asked about MTPE or EOS, but the space agency followed up with responses to written inquiries. Of the eighty-eight questions asked by the committee, the twenty-fourth foreshadowed its thinking:

> Question 24: Given the very real budget constraints the committee faces, both in 1992 and the years beyond, isn't it advisable to redesign the EOS project to reduce its $18 billion budget cost over the next nine years, but hold its current launch schedule? How would you rescope the project if we told you to hold the schedule for launch of the first platform . . . but reduce the program's cost by $6B?
>
> Answer 24: A $6 billion reduction in funding for EOS between now and 2000 would require NASA to significantly delay and re-phase portions of the program. Assuming that the program would be fully funded

at the current projected level in 1992–94, and the suggested reduction was made after 1994, an EOS program could be designed with the following characteristics: development, delivery, and launch of the first EOS-A series spacecraft on schedule for the fourth quarter of 1998; online capability for the prototype EOSDIS (version 0) on schedule for 1994; delay of 2–3 years of the fully functional EOSDIS, which may not be able to support EOS-A in time; therefore we may have to consider a delay of EOS-A; delay all plans for follow-on EOS-A platforms; and a total redesign of the EOS-B concept.

When the Senate finally took action on the NASA budget request, this was precisely what the space agency was faced with regarding EOS implementation.[27]

In March 1991, NASA established the EOS Engineering Review Committee to study options for dramatically reducing the project's scope. The charter for the new task force stated,

> The objectives of the review are to ensure that the recommended program: (1) meets established scientific objectives; (2) places [the] highest priority on achieving data collection needs specifically focused on climate change and global warming issues; (3) minimizes annual funding requirements; (4) minimizes technical risk; and (5) has sufficient resiliency to be adaptable to requirements.
>
> This engineering review will include . . . an analysis of cost, budget, technical, and scientific trade-offs of alternative approaches to the EOS instruments; an analysis of cost, budget, technical, and scientific trade-offs of launching and flying all EOS instruments with a series of large platforms or with a series of smaller platforms; [and] . . . an analysis of NASA's plans to allow for evolution in EOS technology and science goals.

In consultation with the Space Council, Admiral Truly selected Scripps Institute of Oceanography Director Edward Frieman to assemble a panel and work with it to provide recommendations for restructuring EOS. Frieman was a plasma physicist who had served as assistant secretary of the Department of Energy as well as an executive vice president for SAIC. Given his background and the long-term role that Scripps had played in studying atmospheric carbon dioxide, Frieman seemed a good choice to conduct this study.[28]

In late April, Frieman traveled to Washington to meet with Vice President Quayle, Richard Truly, Mark Albrecht, Shelby Tilford, EOS program scientist W. Stanley Wilson, and a number of congressional staffers. He used these interviews to "basically . . . explore the various positions on EOS and why the review was called." He heard the various critiques of the project, including concerns that EOS was too expensive, did not make use of new small satellite technology, and might not be in operation soon enough to respond to critical near-term global change policy concerns. Frieman was surprised by how frequently these criticisms were raised by those whom he interviewed. He found open hostility toward EOS within the Space Council staff, which had been brewing since the start of the Bush administration. Albrecht and others had deep experience with national security space programs, such as the Strategic Defense Initiative, and adamantly recommended that Frieman carefully consider placing EOS sensors on small, single-instrument payloads, which they believed would cost much less. Advocates of this tactic also argued that the importance of simultaneity was overstated but that it could nevertheless be accomplished by maneuvering multiple small satellites into formations capable of achieving the same results. It was claimed that the smaller satellite method would produce research results much sooner, satisfying the critical needs of both scientists and policymakers. Finally, the small satellite approach had the backing of at least one major industry actor. At the same time that Frieman was making his rounds, McDonnell Douglas officials were briefing space agency audiences on the desirability of using the Delta II launch vehicle to carry out these missions. While in Washington, Frieman became increasingly aware of the budgetary uncertainty facing the program. Representative George Brown (D-CA), who chaired the House Committee on Science, Space and Technology, sent him a letter making it clear that "despite my own conviction and that of many of my colleagues that increased investments will be needed to better define future environmental policy options, a major scientific undertaking such as EOS . . . is difficult to sustain in this budgetary environment."[29]

By the time he left Washington, Frieman clearly understood the enormity of the task facing his panel and committed to providing as detailed an assessment as possible within the short amount of time available. In negotiations with NASA, he had ensured its leaders that his study would investigate both the EOS-A and EOS-B platforms despite the efforts of some to limit the scope to the former satellite. In late May, the EOS Engineering Review Committee held its first meeting in Washington. Over the course of two days at NASA headquarters, agency officials, EOS's principal investigators, and non-NASA participants briefed the panel. The committee adjourned after finalizing plans for a detailed workshop in late July at the Scripps Institute of Oceanography, where it expected to make the majority of its tough decisions.[30]

Within days of this initial meeting, the House Appropriations Subcommittee released its markup of the FY1992 budget. Heeding the advice of the Augustine committee, the subcommittee launched an attack on the space agency's flagship program: "The committee deeply regrets that, owing to overall budget constraints, it has been forced to suspend funding for the space station program. This decision is reflective of the funding crisis facing all domestic discretionary programs. . . . It is not possible to fund 10-to-15 percent increases for all federal domestic discretionary agencies when total growth permitted by the 1991 Budget Summit Agreement is less than 7.5 percent." This subcommittee decision kicked off a fast and furious debate within the legislature and the space policy community over whether the space agency should focus on robotic or human exploration. NASA and the White House launched a full-scale effort to paint the decision as one that would signal an American retreat from leadership in human spaceflight. In fact, the reasoning behind the move had been focused on ensuring full funding for the organization's robotic science programs, including EOS, and other independent science agencies. Regardless, bipartisan forces led by the House Science Committee's Bob Walker (R-PA) rallied behind the human spaceflight program. On 6 June, they forced a vote of the full body to amend the subcommittee's markup and reinstate funding for the station. After a seven-hour debate on the chamber floor, the markup was overturned, and a new budget plan for NASA was adopted by a vote of 240 to 173.[31]

The new plan cut the budget for Space Science and Applications by $463 million (16 percent), aeronautical research by $79 million, and shuttle operations by $233.6 million. The House did not specify where Fisk was supposed to make these cuts, but likely targets included the Advanced X-Ray Astrophysics Facility, Cassini, and EOS—all of which would face major delays or outright cancellation as a result. Bob Traxler said of the new budget plan, "Big science is really knocking out little science here. The

Space Station is going to eat your dinner next year." Many in the space science community echoed his concerns. Former JPL Director Bruce Murray said it was a watershed moment and "the end of a set of relationships that go back more than 30 years." Marcia J. Rieke, director of the University of Arizona's Steward Observatory, said it sent "a very depressing message. The scientists in this country have been working hard to do a good job. To have that kind of message come back—that the country doesn't care about what you're doing or its future—is demoralizing in the extreme." The title of a *Washington Post* article by Kathy Sawyer summed up the feeling of the science community: "Space Budget Battle: Humans 1, Robots 0."[32]

With the House action finalized, the focus turned to the Senate. Thomas Donahue, former chair of the Space Studies Board, noted at the time that Barbara Mikulski's support for the human spaceflight program meant that it "would not be productive for [those] in the space science community to wage a battle against the space station at this stage." On 11 July, the Senate subcommittee released its NASA budget markup. The document stated, "The committee has reluctantly [limited] NASA's total increase to just over 3 percent over its [FY1991] appropriations because of the extremely severe fiscal constraints." The subcommittee, however, seemed to strike a balance between human spaceflight and science: "Included within the committee's recommendation are the full amount requested for the Space Station Freedom project, as well as a 10-percent increase for NASA's space science programs. Critics of the space station have suggested that it will cripple space science by diverting funds that would otherwise be available for scientific missions. The Committee's recommendations contradict that criticism." Regardless of this contention, there were clear winners and losers. While the station and space science came through looking very good, MTPE and EOS had lost significant ground.[33]

While the subcommittee approved only a slight cut in the EOS budget for the next fiscal year, it supported much more drastic out-year reductions. The Senate wanted NASA to cut the overall budget by $5 billion and made it clear that the agency had to be prepared to deal with new fiscal realities. The subcommittee's findings are worth quoting at length:

> The committee notes the growing concern within many sectors about the cost, complexity and schedule of the EOS program. Despite these concerns, the committee continues to believe that it is among the most important of NASA's current science projects because its results will

have profound implications for future economic and policy decisions on how to respond to the growing concern over the Earth's climate change. Nevertheless, with a current cost of over $16,000,000,000 between fiscal year 1991 and fiscal year 2000, the committee does not believe that the current budget climate will make allocating this level of funds possible. Therefore, the committee directs NASA to modify the scope and cost of the EOS project, while keeping intact its principal objective of providing a long-term database of information and predictive capability about the Earth's climate. NASA should submit its results to the committee by March 1, 1992.

Several principles should guide the agency in this redesign effort. First, the committee expects NASA to reduce the program costs from the current baseline by approximately $5,000,000,000 between now and the year 2000. Second, NASA should anticipate a fiscal year 1993 budget for EOS that is no greater than $150,000,000 to $200,000,000 above the final fiscal year 1992 appropriation. Third, highest priority should be given to maintaining the current launch schedule of a 1998 launch for the first EOS-A platform. Fourth, sufficient funds should be set aside to guarantee that the system is clearly interdisciplinary in nature. It should include adequate archival and retrieval capability for the EOSDIS portion of EOS, with sufficient reserve to handle unanticipated problems in the development of the EOSDIS system. Fifth, and finally, NASA should carefully examine the feasibility of using smaller platforms, with a smaller total complement of instruments, to accomplish EOS goals and objectives in the context of the current constrained fiscal environment. On this last principle, the committee is not predisposed to a particular outcome. Nevertheless, it is a reasonable area to explore given current fiscal realities.

As a result of these congressional actions, when the Frieman committee met the following month, it had a new charter: to generate a response to this proposed new direction for MTPE.[34]

When the EOS Engineering Review Committee convened in La Jolla, its members were split into two roughly coequal camps. Scientists from the space agency who favored delaying the EOS schedule in response to the proposed budget reductions populated the first group. The second group was made up of outside scientists who favored shrinking the EOS platforms and putting fewer instruments on more satellites. This second group found its interests bolstered by a collection of presentations suggesting that using small satellite technology would result in significant future benefits for earth observations and global change research. James Hansen, from NASA's Goddard Institute for Space Studies, presented a proposal for a small climate satellite called CLIMSAT that could carry three instruments to measure atmospheric aerosols, the earth's radiation flux, and various cloud properties. He suggested that it could be launched by 1997 and would only cost $350 million. Lowell Wood, from Lawrence Livermore National Laboratory, presented a more radical idea called Brilliant Eyes, which would involve a constellation of small, single-instrument satellites flown in place of EOS-A. The JASON Defense Advisory Group recommended that the federal government invest in reducing the cost and size of remote sensing instruments and that the 1990s be used as a period to demonstrate these advanced technologies. Researchers from the Defense Advanced Research Projects Agency (DARPA) and Sandia National Laboratory also presented ideas for small satellite projects.[35]

While the Frieman committee members were initially predisposed to maintaining the broadest possible science objectives for EOS, Jack Fellows of the OMB and Colonel Pete Worden (USAF) of the National Space Council informed them that the administration would be able to provide support only for a more narrowly tailored program. The panel agreed that this new focus should be on climate change and decided to use Intergovernmental Panel on Climate Change (IPCC) recommendations to guide the modified program. The 1990 IPCC report had called for an increased emphasis on various climate-related processes, particularly those associated with clouds, oceans, and the carbon cycle. These three priorities became the basis of EOS's scientific refocusing. Frieman later said that once the focus of EOS shifted, "the driving force of having the original 18 or so sensors on one large platform [didn't need to] be maintained." As a result, the committee concluded that simultaneity could be achieved to meet climate science requirements by employing smaller clusters of sensors on one satellite.[36]

During the EOS Engineering Review Committee meetings, the panel learned from officials at General Dynamics that the USAF had decided to modify pad SLC-3E at Vandenberg Air Force Base to launch the Atlas IIAS. This was confirmed during a USAF presentation. As Frieman later recalled, "The other piece of information which came forth during the course of our deliberations [was] an Air Force requirement for the Atlas

IIAS. We [were] told repeatedly that this was a hard requirement. We [were] told that there [was] money in the budget. . . . The information provided to us said that this option would be available at the Western Test Range beginning in 1995." This new information changed the nature of the decisions that the panel faced, particularly with regard to the fate of the EOS-B platform. If NASA could leverage the Atlas IIAS capability, it would be able to launch small and intermediate satellites from the West Coast into polar orbits. Combined with the announced budget cuts and new focus on climate change, this "dictated in a sense [the committee's] first major finding: that both EOS-A and EOS-B can and should be completely reconfigured." By this time, the group already favored breaking the EOS-B spacecraft into a smaller set of satellites, and the last few days of the meeting were used to develop guiding principles for EOS's restructuring. This work continued through August, and on 11 September, the Frieman committee report was completed and delivered to NASA.[37]

The report laid out several notional architectures based around small spacecraft options, noting that EOS-A and EOS-B instruments "should no longer be thought of as separate packages." The panel recommended an approach with three medium-sized spacecraft launched aboard an Atlas IIAS instead of two large spacecraft launched aboard a Titan. The committee wrote,

> This Atlas option contains a favorable measure of resilience over the Titan options [because it] falls below the (Senate) budget mark used for FY92 through FY94. It should be noted that this is accomplished by delaying or deleting EOS-A instruments that were included in the previous Titan cases. . . . The committee recommends that NASA use the Atlas IIAS concept discussed here as a proof of concept, and explore the details of instrument selection and flight with the relevant scientific communities.

In suggesting a change to the EOS architecture, the Frieman committee realized that a corresponding change to EOSDIS was also required. The panel supported scaling down EOSDIS to a size more compatible with the new approach, but it also wanted to ensure that the space agency would be able to handle the massive amounts of data that would still be coming from EOS. Therefore, it recommended that "NASA reexamine EOSDIS both as to its utility in a redesigned program and on its merits in handling the data and information needs of such a large component of the [USGCRP]."[38]

EOS Engineering Review Committee recommendations significantly reduced the scope of the EOS science mission, most notably in the areas of stratospheric chemistry and solid earth physics. Frieman admitted as much when he briefed the House Science Committee about the results of the study: "All we have said and are trying to maintain is that those highest priority science issues raised by the [IPCC] will be addressed under this program. But I do not want for one moment to say that all of the science that was originally [packaged] into the NASA proposal is going to be addressed." Nonetheless, he expressed his belief that the committee had made an appropriate recommendation to alleviate the many concerns expressed about EOS during the previous year and a half. It was left to NASA to follow through.[39]

On 27 September, a House-Senate conference committee finally agreed on NASA's FY1992 budget. Following the Senate lead, the conferees cut the agency's budget request for EOS by $65 million (from $336 million to $271 million). They did, however, provide the full request of $82 million for EOSDIS and increased the earth probes line by $20 million. The conference report stated,

> The conferees concur [with] the Senate language that directs the restructuring of the EOS program with the following adjustments. First, the program is capped through fiscal year 2000 at $11,000,000,000. Second, NASA is directed to place the EOS instrument configuration on smaller, multiple platforms rather than the single platform approach proposed in the fiscal year 1992 budget submission. NASA should assess the various launch option configurations available before recommending a final framework to the committees. Finally, NASA should submit its restructured EOS program to the committees for their approval by February 1, 1992. NASA should refrain from any additional contract awards for any instruments until the Committees on Appropriations have approved the restructuring plan.

NASA spent the next several months working on a redesign plan in response to the congressional direction. A three-day meeting on 21–24 October, which was attended by program scientists and principal investigators, gave the space agency new recommendations with regard to specific instrument pairings and spacecraft phasing. From 21 to 27 November, NASA planners led by Chris Scolese, Marty Donohue, and Dixon Butler finalized their specific plan in a long series of meetings held in La Jolla, California. On 11 December, Shelby Tilford and Len Fisk briefed Truly, and the administrator endorsed the suggested changes and ordered that they be incorporated into the agency's FY1993 budget for negotiation with the Bush White House. The OMB agreed to include $391 for EOS in the president's budget (reducing the agency submission by $20 million) but maintained the total request of $82 million for EOSDIS and $88.9 million for the earth probe missions.[40]

As had been recommended by the Frieman committee, Administrator Truly sent a letter to NRC President Frank Press requesting that the

> Academy undertake a high level review of [NASA's] plans for the EOS Data and Information System (EOSDIS) to ensure that this effort is compatible with the restructured [EOS] program and best serves the interests of a broad range of users. . . . NASA is determined to proceed as rapidly as possible with the restructuring of the EOS program in order that there will be no unnecessary delays in implementation. It is important that your study be accomplished such that an interim or status report can be available to us by March 1992 and a final report of your findings by summer 1992.[41]

After appointing Charles Zraket of Harvard University to chair this committee, NRC leaders identified three chief tasks for the panel: first, to determine whether EOSDIS was being developed to meet the needs of the science community; second, to determine whether NASA was managing EOSDIS properly, and specifically whether the system could be operational in time for the launch of EOS AM 1; and finally, to determine whether EOSDIS would be flexible enough to meet the evolving needs of its user community. The Zraket committee began meeting in mid-February 1992 with the goal of presenting NASA with an interim review within a few months.

On 26 February 1992, Len Fisk presented the newly restructured EOS at a hearing before the Senate Subcommittee for Science, Technology and Space. He opened by downplaying the recent struggles with the Frieman committee and suggested that "in lieu of the budget cuts and the appearance of the Atlas IIAS, NASA would have likely gone to this kind of configuration without the Engineering Review."[42] He revealed that rather than pursuing the broad-based science objectives previously imagined for the EOS-A and EOS-B series, the new satellites would focus on climate change research and seek to better understand:

- The role of clouds, radiation, water vapor, and precipitation.
- The productivity of oceans, their circulation, and air-sea exchange.
- The sources and sinks of greenhouse gases and their atmospheric transformations.
- Changes in land use, land cover, primary productivity, and the water cycle.
- The role of polar ice sheets and sea level.
- The role of ozone chemistry.
- The role of volcanoes.

NASA had abandoned plans to study geomagnetism, solid earth geophysics, and some aspects of upper-atmosphere chemistry. The largest instruments, including a space-based large-aperture lidar and synthetic aperture radar, were dropped because they were too large to be accommodated on the much smaller Atlas-sized satellites. A written report that NASA presented to Congress a couple of weeks later stated, "The restructured EOS program, including the spacecraft, EOSDIS, and the supporting science, has been designed to accommodate the $11 billion funding cap through 2000. . . . The total cost of the space hardware component, through FY2000, is approximately $6 billion. The ground segment—EOSDIS and science—totals to $5 billion. This proportion of ground-to-flight segments is very high by the standards of any previous NASA program."[43]

EOS now consisted of six payloads—three Atlas-sized satellites, one Delta-sized satellite, and two satellites sized for a new small launch vehicle intended to replace the discontinued Scout rocket. The first Atlas-sized satellite, *EOS-AM*, would carry instruments to measure the terrestrial surface, clouds, aerosols, and the earth's radiation balance—it was a polar-orbiting sun-synchronous satellite that would cross the equator in the morning. The second Atlas-sized satellite, *EOS-PM*, would carry instruments dedicated to measuring clouds, precipitation, terrestrial snow, sea ice, sea-surface temperatures, and ocean productivity—it was another polar-orbiting sun-synchronous satellite that would cross the equator in the same location as *EOS-AM* twelve hours later. The third Atlas-sized satellite, *EOS-CHEM*, would carry instruments to measure ozone and atmospheric chemistry (replacing the UARS)—it was also a polar-orbiting sun-synchronous satellite. Each spacecraft had a planned five-year lifetime, and it was expected that they would be replaced twice (for a fifteen-year coverage period). No final decision had been made, however, regarding what instruments would fly on the follow-up platforms.[44]

The Delta-sized payload, *EOS-ALT*, would carry instruments to measure ocean circulation and ice sheets. The two smallest satellites, inspired by Lowell Woods's Brilliant Eyes concept, were single-instrument platforms named *EOS-COLOR* and *EOS-AERO*. *EOS-COLOR* would fly in formation with *EOS-AM* and measure ocean color, while *EOS-AERO* would fly in formation with *EOS-PM* and measure atmospheric aerosols—combined, the two satellites were intended to provide data needed to improve models of atmospheric dynamics.[45]

The restructured program still did not completely silence criticism from within the scientific community. Several points of contention remained, including the fact that there had been no schedule improvements. The most important point of controversy, however, was the launch order for *EOS-AM* and *EOS–PM*. Many scientists believed that *EOS-PM*'s instruments were most relevant to detecting and measuring climate change and therefore argued that it should be launched first. The primary counterargument was that *EOS-AM* was already under construction, and delaying it would increase the overall mission costs. Additionally, the most expensive instrument aboard this spacecraft (the Advanced Spaceborne Thermal Emission and Reflection Radiometer [ASTER], a high-resolution land imager) was being provided by Japan, which dramatically reduced early-year budget requirements. Finally, *EOS-PM* was considered much more challenging to implement because of the technological complexity of its flagship instrument (the Atmospheric Infrared Sounder [AIRS], which would use infrared technology to create three-dimensional maps of air and surface temperature, water vapor, and cloud properties). The EOS Program Office did not believe that the AIRS development could be accelerated successfully; thus, *EOS-AM* appeared to be the better choice for the first program launch. Fisk and Tilford stood firm in their belief that nothing would be gained by trying to rearrange the flights.[46]

❧ ❧ ❧

When it was initially conceived, EOSDIS was modeled on large-scale DOD computing projects that used a centralized data processing concept characterized by large and expensive high-powered platforms and a limited supply of programming talent during the early years of digital computing. For EOSDIS, this centralized model would be manifested in a large-scale computing installation that would provide satellite command and control, develop the instrument retrieval algorithms, process and validate data sets, and then provide them to offsite Distributed Active Archive Centers. As the Zraket panel began meeting to address these issues, it became increasingly clear that such centralization of computing resources was no longer necessary. Instead, the study team suggested that a distributed system made much more sense.[47]

The panel contended that the centralized EOSDIS design did not have the required flexibility to achieve the desired EOS results. The major problem was that this concept was based upon an automated data distribution system that would provide a set of standardized products to users when what researchers really needed was the ability to "combine data from different sensors, alter the nature of the products to meet new scientific needs, or revise the algorithms used to process data for different purposes." The ability to perform this sort of interdisciplinary investigation had been one of the major selling points for EOS, but the study team did not believe that existing plans for EOSDIS would support this objective. Furthermore, the growing availability of computing power made a distributed architecture both possible and desirable. This new distributed architecture model would be based on the ARPA network (ARPANET), which the following year would become publicly available as the Internet. ARPANET had already demonstrated that a distributed computer network could change with extraordinary speed as user demands shifted.[48]

Therefore, the Zraket committee recommended redesigning EOSDIS around the Distributed Active Archive Centers, which would act as the system's interface with the EOS user community (while an "EOSDIS Core System" would provide the network infrastructure required by the archive centers). The panel's vision was that EOSDIS would ultimately evolve into UserDIS, with science users able to access data sets from multiple sources and integrate them into new products. This meant, in turn, relying on the "entrepreneurial spirit of the [Distributed Active Archive Centers] and other interested organizations." In the end, the belief was that this approach would result in a much more flexible data system capable of adapting to new uses and new user-generated products as EOS itself evolved. As NASA moved forward based on these recommendations, as well as the Frieman committee suggestions for restructuring EOS, it was hoped that the space agency would be able to proceed in the coming years without further changes to the program. Those hopes were completely dashed over the coming six months, and the program would in fact remain under fire for an additional four years. Luckily, the agency had the support of the science community during these ordeals, especially because they forced NASA to make a series of pragmatic choices that would essentially save the entire initiative.[49]

## CHAPTER FIVE

# Last Hard Push to Secure EOS

In fall 1991, as the Frieman committee was working to restructure EOS, the world had begun to shift under NASA's feet once again. The initial stimulus for this upheaval was the decision by NASA Deputy Administrator J. R. Thompson to tender his resignation. The Space Council staff believed this was an opportunity to select someone who would wholeheartedly support President Bush's vision for the future. When Mark Albrecht began his search for a replacement, however, he was surprised to find that no one would take the position as long as Admiral Truly remained administrator. In December, Vice President Quayle and Albrecht met with three former NASA administrators to discuss this dilemma. James Beggs, Thomas Paine, and James Fletcher all reiterated a common message—Truly had to go.[1]

Quayle transmitted this message to President Bush, who approved of Truly's removal and suggested that he be appointed to an open ambassadorship. Quayle summoned Truly to the White House and requested that he step aside. The administrator said he would consider the proposal but within a few days sent a letter to Quayle saying he would not resign. Over the course of the next couple of months, Truly made an effort to keep his position and demanded an opportunity to plead his case directly to President Bush. On 10 February 1992, this meeting was arranged, and after a half-hour discussion, Truly finally agreed to submit his resignation.[2]

There were mixed reactions regarding the decision to fire Admiral Truly. The George Washington University's John M. Logsdon, who was a member of the vice president's Space Advisory Board, was quoted as saying that Truly "did an extremely valuable job in getting the Shuttles flying again, and restoring a sense of integrity to the agency, [however], Truly's vision of the future was not compatible with the realities of the world." Senator Al Gore was more critical of the administration: "I view this as a very troubling sign that . . . Quayle's Space Council may have forced Admiral Truly to leave this job because of the [Space Council's] insistence on running NASA from the Vice President's office."[3]

President Bush put Mark Albrecht in charge of finding a replacement for Truly. The latter's resignation would be effective on 1 April, which meant that Albrecht had only forty-five days—this would be the fastest confirmation process the Bush administration had ever faced. Within a few days, Albrecht had compiled a short list of potential candidates, filled mostly with well-known names such as Norman Augustine. But one name that stood out was that of Daniel Goldin, a relatively obscure middle manager at TRW. Goldin was a mechanical engineer who had received his BS from City College of New York in 1962; his first job after graduating was at NASA's Lewis Research Center, where he dreamed of sending humans to Mars. Within five years, he was at TRW working on classified defense programs, and during the mid-1980s, he led many projects for the Strategic Defense Initiative Office. As Henry Lambright notes, "in the 'black' world of military space Goldin was a rising star with a reputation as a hard driving innovator. He, in particular, was winning attention by using very advanced microelectronic technology to launch smaller spacecraft."[4]

Goldin believed that NASA was on the wrong track and was exhibiting the signs of bureaucratic aging, which he believed particularly afflicted the agency's science programs. He spoke of a "vicious cycle" afflicting OSSA and other NASA entities: NASA loaded a large number of experiments onto a few big, expensive machines that were launched into space. The scale of the enterprise meant that it took a long time to get the spacecraft developed and operating. Because it took so long to get these spacecraft built, they incorporated obsolete technology by the time they reached orbit. With so much incorporated into these expensive machines, NASA could not afford to lose any of them. The agency had become risk averse, he said, and emphasized extrareliable (i.e., less innovative) technology. If anything ever did go wrong, NASA took a huge political hit because so much money and time appeared to have been wasted.

The Space Council staff had taken note of Goldin a couple years earlier when he had made a pitch for a smaller, cheaper version of EOS. He was part of a crowd who thought EOS was unnecessarily large and expensive—he referred to the platforms as a "Battlestar Galactica." When he had pitched his ideas to NASA officials, they had threatened to cut off contracts with TRW if he pursued them. When his name surfaced for consideration as administrator, he relayed this story to Albrecht—which probably boosted his candidacy.[5]

Considering that President Bush was already in the midst of his reelection campaign, one worry was that no one would be willing to become NASA administrator. But when the position was tendered, Goldin was flattered by the presidential offer and accepted the nomination. He was excited about an opportunity for public service and wanted to bring more innovation to U.S. technology policy. The White House was confident that his nomination would sail because, in addition to his qualifications and personal charisma, he was also a registered Democrat. On 11 March, the White House announced Goldin as its nominee to become the next NASA administrator.[6]

The response to Dan Goldin on Capitol Hill was extremely positive. During his Senate confirmation hearing, he communicated what he called his four core initiatives for NASA: a human presence in space to understand the space environment and make progress to eventually explore in a more effective, long-term way; understanding what is happening on Earth, enabling policy makers to manage by fact, not by intuition; understanding the universe—what it is made of, how it formed, and what is happening to it; and maintaining aeronautics leadership. He also said in his opening statement, "I will be in charge of NASA." This comment resonated with Senator Gore, who remarked during the hearing about his displeasure with the Space Council: "You and I talked privately, and you know that NASA, as an institution, is now faced with this problem of the National Space Council expanding its role from what some regard as a quite legitimate role in looking at policy. . . . Interference in the management decisions of the NASA administrator crosses the line." Goldin reassured Gore that he would be in charge, and Gore responded, "I sense a backbone in this nominee, Mr. Chairman." Goldin was approved overwhelmingly and was officially sworn in during a brief Oval Office ceremony on the afternoon of 11 April.[7]

With the arrival of a new administrator at NASA, plans to restructure EOS were essentially null and void within three months of Len Fisk briefing Congress regarding them. In the first month of his tenure, Goldin traveled to all ten NASA field centers to meet with senior officials and learn about their work. During his tour, he saw many things that he did not like about the agency, including inadequate cost and engineering management, excessive reliance on outside contractors, a lack of consistent and successful management practices, and poorly defined roles for individual project managers. Having completed his tour and identified these problems, Goldin immediately began working to implement a new technical and management philosophy that he called *faster, better, cheaper.*[8]

From a technical perspective, this new approach relied quite literally on new innovations to reduce spacecraft size and weight. For NASA, this meant leveraging capabilities developed within the national security space program. From a managerial perspective, this new approach relied on smaller project teams and shorter schedules to complete a mission. Advocates of the philosophy believed that hard work and detailed attention to procedures would result in projects that would meet the same science and technology standards as larger, more expensive spacecraft. The combination of advanced microtechnology, smaller teams, faster schedules, and the expectation of equal output meant that programs would have to accept more risk. At the time of Goldin's appointment, there was no consensus opinion that *faster, better, cheaper* would actually work for NASA.

After Goldin's travels, he directed Acting Deputy Administrator Charles Bolden to prepare NASA's senior leadership for a bottom-up review of all agency programs. On 7 May, Bolden sent a memorandum to all associate administrators and field center directors regarding the objectives of this review:

> To formulate a shared vision for NASA, which will encourage [the] revitalization of our industry, lift the spirit of pride of America and the NASA team, and ensure U.S. leadership in space and aeronautics. . . . To assess our present ways of doing business and evaluate alternative approaches that can improve management, increase efficiency, save money, enhance schedules, and reduce risk. . . . To evaluate options to current programs that will support development of balanced and complementary plans for NASA and consider the essential elements of NASA's commitments to its stakeholders, yet be consistent with budget realities.
>
> . . . In order to tackle the [above tasks], we will employ a concept through which [Administrator Goldin] experienced great success in the corporate world. This concept involves using two teams (e.g., Red and Blue) tasked to conduct simultaneous planning and evaluation of pro-

grams/projects. Our Blue teams will be made up of the program and center personnel who will be conducting critical bottom-up, internal reviews of their respective organizations. The Red teams will be [made up of] seven mission review panels and . . . institutional and special initiative teams. [The] major emphasis of the Red teams will be to bring independent, critical thought and critique to . . . associate administrators and center directors as an integral part of the review process—to provide fresh ideas to the associate administrators and center directors! In simple terms, the Blue teams will have prime responsibility for the poking, probing, and self evaluation; the Red teams will be the conscience and sanity checkers to ensure no fresh ideas get stifled. Through those checks and balances, we will focus our ideas into a shared vision and an integrated plan to make that vision [a] reality.

Among the seven mission review panels established was one for Remote Sensing and Environmental Monitoring of Planet Earth.[9]

In addition to his concerns about NASA's internal management and personnel problems, Goldin did not believe the space agency was going to receive the funding it needed to support all the projects on its plate. He said at the time, "When I took a look at the run-out budget, I became terribly, terribly concerned. America demands we produce for much less, or NASA will drift into irrelevance." As a target for achieving the kind of reform he was looking for, Goldin told his blue and red teams that he wanted them to find 30 percent savings in every program or project they reviewed. This was intended to force the teams to "strive to examine fundamental changes in programs, not just fine tune them to save some money." These cuts, however, would not necessarily be applied across the board. It was a straw-man approach: "We're not going to take precipitous action. We are going to preserve stability in NASA. I am asking us to step back, catch our breath, figure out where we're going, come up with a plan and then say [what] we ought to do about it."[10]

Goldin gave the blue teams an aggressive schedule, asking them to complete their internal reviews with their red team counterparts by 15 June. Four days later, they were expected to present a plan to an integration team led by Michael Griffin, associate administrator for exploration. Five days after that, Goldin would be briefed about the findings. The objective was to have final decisions made by 1 July, with an eye to making necessary changes to the agency's FY1994 budget request.[11]

Per Goldin's instructions, the EOS program began its "rescoping" exercise in an effort to find 30 percent (approximately $3 billion) in potential budget cuts. Christopher Scolese, a project engineer from the Goddard who had also been a critical player in the recently completed restructuring exercise, led the blue team. On 20 May, the team met for the first time and set about determining how to implement budget cuts while minimally impacting the science mission and maintaining a focus on the IPCC priorities. It did not consider constraints with regard to launch vehicle selection, spacecraft lifetime, technical requirements, instrument pairings, or existing contractual commitments.[12]

The Scolese team recommended maintaining the status quo for *EOS-AM 1* but suggested that all other spacecraft should be changed. It maintained the scientific focus of each set of spacecraft as defined by the restructuring exercise, but all spacecraft after *EOS-AM 1* would be built on a common spacecraft bus—all sized for launch on an intermediate launch vehicle (these smaller vehicles would be less expensive). The group urged that contingency reserve funding (extra funding for unanticipated problems during development) be reduced for both *EOS-AM* and *EOS-PM* by several hundred million dollars. The team eliminated the *EOS-AERO* spacecraft from the budget altogether. Another important deletion was the High-Resolution Imaging Spectrometer (designed to acquire images in 192 spectral bands simultaneously), an expensive instrument that was budgeted at nearly $500 million and had been intended to ultimately replace instruments on *EOS-AM 1* and *EOS-PM 1*. Its elimination allowed the team to fix the configurations of the *EOS-AM 1* and *EOS-PM 1* follow-up spacecraft, saving significant development funding. The team reduced the size of the AIRS instrument (designed to observe and characterize the entire atmospheric column from the surface to the top of the atmosphere) and assumed a build-to-cost approach for all remaining NASA instruments, which would act as a built-in cost cap. These and other decisions were based on the belief that international partners would independently launch similar instruments. Finally, some instruments were shuffled around between different satellites, all with an eye toward making the entire fleet of spacecraft smaller, similarly sized, and significantly cheaper.[13]

In addition to recommending changes to different EOS platforms, the blue team also advocated altering the EOSDIS architecture. Most importantly, it was suggested that to cut costs, the number of data products available at the time of the *EOS-AM 1* should be reduced. This would require that the system focus on only the highest-priority science. In doing so, the team anticipated the EOS program implementing the Zraket committee's recommendations even though there would be some costs associated with changing the EOSDIS structural design. To balance the need for these outlays, the panel recommended reducing the system's processing capacity by more than half and the number of data products by nearly eightfold (from just over eight hundred to just over one hundred). It was believed the reduction in data products accomplished by transitioning to a distributed architecture would not have significant negative impacts because outside researchers were expected to use the available data sets to create their own data products. In fact, many in the scientific community believed that this innovation would be one of the primary benefits of a decentralized network. What was lost was the ability to process huge amounts of data in real time. Instead, data releases would be made months after receipt. This violated the original intent of EOS but was not perceived as a great loss because it had never been a part of previous earth science missions.[14]

These EOSDIS recommendations were controversial because the blue/red teams made them without consulting with the EOS Payload Advisory Panel. After the completion of the reviews, Shelby Tilford asked the panel to review the suggestions and the proposed rescoping of EOS. The panel almost unanimously agreed with the proposed rescoping, but it noted, "The Red/Blue team has started the process of refining the list of science data products to be provided in EOSDIS. However, the EOS investigators cannot relinquish responsibility for this task. Hence the science panels and instrument teams of the EOS Investigators Working Group must systematically develop the list of core data products, including science requirements, algorithm heritage, alternative approaches, and intermediate products." The EOS Program Office agreed to work with the advisory panels to determine which data products would be of the highest priority, but it did not back away from its plan to reduce the output of EOSDIS.[15]

By July, the blue team review was essentially complete, and NASA began internal preparations for submitting its FY1994 budget request based on the assumption of a rescoped EOS—with a total budget of $8 billion through the end of the decade. The following month, the Senate Appropriations Committee released its markup of NASA's FY1993 budget. The document contained some good news (the agency received its full $391 million request for EOS and EOSDIS) and hewed closely to what had just been decided by Scolese and his colleagues:

> Future restraints on Federal spending make it likely that the fully restructured baseline for the EOS project, $11,000,000,000 through fiscal year 2000, is not attainable. As a result, the Committee directs NASA to adjust the EOS program in the following manner, contingent on full funding for the program in fiscal year 1993. First, the total project cost through fiscal year 2000, exclusive of construction of facility, launch, and tracking requirements, shall be capped at $8,000,000,000. The Committee considers this amount to be a new funding floor below which the project shall not go, and has included bill language for that purpose. Second, the agency shall submit, concurrently with the annual operating plan, a firm, fixed cap on the development costs for all EOS instruments through fiscal year 2000, with a cap on the amount of funds for instruments specified by each EOS platform, including EOS AM-1. Third, the agency should adopt a common spacecraft approach for all EOS platforms after the initial EOS AM-1 spacecraft. And, [NASA] should establish a management plan for its network of eight data active archive centers (DAACs), including a precise delineation of the scientific and policy roles for each DAAC.

On 24 September, the budget conference adopted this language and reduced the EOS budget to $8 billion through 2000.[16]

That fall, Dan Goldin announced that he was reorganizing NASA headquarters. Among the changes, OSSA was broken into three separate offices—an Office of Space Science, an Office of Mission to Planet Earth, and an Office of Space and Life Sciences. The new position of NASA chief scientist was created, and Goldin gave this job to Len Fisk as part of a larger strategy to change the direction of the agency's space science mission. Goldin believed that Fisk was too resistant to change. This was particularly a problem because he was devoted to plans for large EOS spacecraft and was unwilling to pursue small satellite options as an alternative. The organizational shake-up was Goldin "making a statement about power—his power vis-à-vis that of other senior officials." The new administrator's future at NASA was uncertain, given the upcoming presidential election, and he believed that there was a desire among "senior officials to wait him out, figuring he would be gone." But Goldin was in charge, and it was his prerogative to make changes as he saw fit. Within a short period of time, Fisk left the agency.[17]

Shelby Tilford became the acting associate administrator for MTPE, but he too was resistant to changing EOS. The reason was not necessarily that he thought big satellites were unquestionably the right way to go but that he believed the larger platforms provided opportunities to pursue a wider range of scientific inquiries. The reality, however, was that as the EOS budget shrank, spacecraft size and the scientific mission would follow. Tilford had been centrally involved in earth science debates going back to the System-Z days. He had been instrumental in establishing the Bretherton committee, the Committee on Earth Sciences, and the USGCRP. Based on this long-term dedication to ensuring that the government would adopt an aggressive earth science research program, it was very difficult for him to see the space agency backing away from that commitment. As a result, he submitted his resignation to Administrator Goldin later that year. Nevertheless, during the final few months of the year, the situation within the agency settled down a bit. NASA presented its blue/red team results to the OMB along with its proposed FY1994 budget—which provided $505 million for EOS (including EOSDIS). This rescoped request was fully adopted by the White House, and once again it seemed that the program was on a solid footing. Unfortunately, there would be more angst in the near future.

## Clinton's Election Leads to Continued Change for EOS

In November 1992, President George H. W. Bush lost his reelection bid to Arkansas Governor Bill Clinton. In contrast to Bush's priorities, it quickly became clear that the American space program would not be a major issue on President Clinton's agenda. Within weeks of taking office, he disbanded the National Space Council and tasked Vice President Al Gore with supervising national space policy. Gore had been very impressed with Dan Goldin when the two had met earlier in the year. Perhaps more importantly, he believed *faster, better, cheaper* was a good example of the kind of "do more with less" approach that the new administration wanted to foster. As a result, Vice President Gore decided to keep Administrator Goldin at NASA, making him the highest-ranking Bush appointee to remain in place under the new administration.[18]

When President Clinton took office, his highest priority was an economic package designed to create jobs and cut the federal budget deficit by approximately $500 billion over five years. In addition to new taxes, the new president intended to close this gap by seeking reductions in discretionary spending. Over the course of the next several months, the administration worked with Congress to put together a package. In early February 1993, as this process got under way, Administrator Goldin was summoned to the White House. During a meeting with OMB Director Leon Panetta, he was informed that the funding for the space agency would be cut by 20 percent over the next five years. As a result, Panetta said there was no alternative but to kill the space station. Before the meeting ended, however, Goldin was able to convince Panetta to give him a few days to develop a working budget that would maintain that program—quite a feat considering that Panetta had long been an opponent of the project. Goldin was convinced that without the space station, NASA's human spaceflight program had no future. Over the course of a weekend, the administrator convened a seventy-two-hour crisis meeting that developed three alternatives that could be pursued within the new budget environment. The following Monday, Goldin used a collection of Lego building blocks to build primitive models of Plan A and Plan B and a single cardboard toilet-paper holder for Plan C. That Tuesday, he used the mock-ups at a briefing for President Clinton's senior staff. He was pleasantly surprised at the end of the meeting to receive the go-ahead to fully develop the three new options within ninety days in an emergency redesign effort. The space station eventually avoided cancellation, although its budget was slashed by $7 billion over five years. What would happen to other agency programs?[19]

In June, the House appropriations subcommittee overseeing the space agency released NASA's budget markup. It cut the space agency's FY1994 request by $700 million (although this was still a $240 million increase over the FY1993 budget). Only $15 million was cut from the budget for EOS/EOSDIS. Although the FY1994 budget process was ongoing, NASA was already planning its FY1995 budget submission for the OMB. Later in June, the OMB ordered the agency to reduce that budget request by 10 percent. Regardless, during its program review two months later, the EOS Program Office proposed a slight increase for its FY1995 budget (with additional small increases in the out-years). This was a reaction to what were considered overly large previous reductions in contingency funding for *EOS-AM 1* development. On 1 September, NASA headquarters decided to submit this request for OMB consideration as part of the full agency budget, thus kicking off four months of pointed negotiations between the two organizations.[20]

As this process was ongoing, Senate appropriators released their markup of the space agency's FY1994 budget request. Although they also reduced the NASA budget by several hundred million dollars, they did little to disturb agency plans for a rescoped EOS. In fact, the subcommittee's report stated its commitment to the rescoped EOS plan:

> The Committee expects the Agency to continue to abide by the principles outlined for this program in the Senate Report 102-356. No steps are to be taken to restructure or rescope the program in any manner that is inconsistent with either the guidance in that report or the Committee's fiscal year 1994 report. . . . Frankly, the Committee wishes to restate what it said about EOS last year: NASA cannot be expected to have EOS be the sole focus of the U.S. Global Change Program. It is not designed to answer every question ever asked about global change, nor should anyone insist that it assume added roles in the U.S. Global Change Program, like extensive analysis of the human dimension of climate change.

On 4 October, the House-Senate conference accepted this Senate language and directed NASA to maintain current EOS budget levels.[21]

Regardless of congressional support, when the White House made its final budget decisions just before Christmas, the OMB had trimmed an additional $758 million from the EOS budget through the end of the decade. The total NASA budget would be $14.25 billion, which was $300 million less than the FY1994 budget. Among the other casualties were all studies and planning for human exploration beyond low-Earth orbit, which effectively erased all remnants of President Bush's SEI. The five-year plan would reduce the agency's budget to $12.6 billion by the end of the decade, with large cuts being made to NASA's human spaceflight programs and aeronautics programs. The overall science budget would actually increase slightly.[22]

Although the overall MTPE budget was cut by over three-quarters of a billion dollars, the situation was actually a bit worse. In its FY1994 budget, NASA had assumed that responsibility for developing *Landsat 7* would be shared with the DOD. However, facing its own fiscal challenges, the DOD had decided to back out of the project. As a result, the White House had delegated full responsibility for the mission to the space agency, which had not been accounted for in the agency's FY1995 budget request. Therefore, NASA would be faced with addressing both the $758 million budget reduction and $300 million in new expenditures for Landsat. Thus, the effective budget hit was north of $1 billion.[23]

Shelby Tilford's departure from NASA left a large hole within the agency and placed Administrator Goldin in a tough spot. It would be hard to find a replacement who was both admired by his or her peers and capable of navigating the institutional minefield between NASA headquarters, NASA field centers, and the scientific community. Tilford had enjoyed a great deal of credibility, even among those who criticized MTPE for political reasons. In August 1993, Goldin began to formally search for a replacement. Among those he spoke with was Ed Frieman, but Frieman ultimately was not interested in the position. Instead, he recommended one of his former graduate students. An astrophysicist educated at Harvard and Princeton, Charles Kennel was a professor at the University of California, Los Angeles and had chaired several NRC Space Studies Board committees. A few weeks later, Goldin called Kennel, and the two agreed to meet in Washington. Three years earlier the latter had participated in a study for NASA's Space Physics Division and had fought against what he called "the trend toward gigantism in NASA experiments." This background greatly impressed Goldin, who decided to offer Kennel the position. However, there was a problem: Kennel was not an earth scientist. Despite his impressive credentials, he was an unknown quantity. He had not been in the MTPE/EOS trenches. In the end, Goldin came to view this outsider status as a positive attribute. He believed Kennel was the kind of change agent that was needed. Goldin told Kennel, "You are going to have to make some very tough decisions about downsizing the Earth Observing System. It will be easier for you if you don't have to do it for colleagues that you've known for thirty years." While there might be problems within the space agency, Goldin believed Kennel would bring credibility to the political battles against MTPE/EOS critics who did not trust NASA's objectivity. Kennel would be seen as someone who was exceptionally competent but was not beholden to anyone. He would reinforce the idea that "science [was] in charge of Mission to Planet Earth." Excited to have Goldin's enthusiastic support, Kennel accepted the job.[24]

From the outset, Kennel faced a bumpy road. The first step was gaining

White House approval, which meant obtaining Al Gore's endorsement. The vice president was barraged by criticism from those who opposed the appointment. Kennel recalled later that his opponents essentially told Gore that the job was "too important to be left to an amateur." Kennel, however, was proactive. He and Goldin contacted those who had objections and informed them that they planned to name Bob Harriss, a highly regarded earth scientist, as MTPE chief scientist. They provided assurances that there would be no filtering out of key science issues. In November, Kennel also met personally with Gore. The two talked about the difficult balance between meeting MTPE/EOS science objectives and responding to a challenging fiscal environment. Gore was sold. On 6 January 1994, Kennel began his tenure as associate administrator and went directly to work to figure out how the space agency should respond to the $758 million EOS budget cut that had been finalized just two weeks earlier.[25]

Kennel started by reaching out to John Klineberg, director of the Goddard, and Berrien Moore, chair of the EOS Payload Advisory Panel, to assess what changes needed to be made. Based on these discussions, it was decided to form three teams to evaluate options for responding to the budget cuts and hidden Landsat costs—a Science Team, a Project Team, and a Review Team. In mid-July, after a whirlwind study, the teams briefed Moore and his advisory panel. A total of twenty-six items were discussed, ranging from outright project cancellations to instrument changes to schedule delays. Each was discussed in a public forum that included EOS investigators and EOSDIS data center representatives. Each session was followed by a private review by the Payload Advisory Panel, where four interdisciplinary subpanels assessed each option. This led to the creation of a new baseline for the EOS program that met high-priority science needs and fit budget constraints.[26]

The following month, the EOS Payload Advisory Panel submitted its report for a rebaselined program for Kennel's consideration. The report contained the following recommendations: make no changes to *EOS-AM 1*; cancel the *EOS-COLOR* mission and instead use *Landsat 7* data; move the *EOS-CHEM 1* schedule by nine months; shift sensors between *EOS-AM 1* and *EOS-CHEM 1*, allowing potential *Landsat 8* instruments to fly on *EOS-AM 2* to save future development costs; divide the *EOS-ALT* mission into two separate spacecraft, placing some instruments on international flights if possible; combine several *EOS-AM 1* and *Landsat 7* science teams to decrease the total number of personnel; decrease the EOSDIS data archiving budget; and expand the planned mission lifetime of *EOS-AM 1*, *EOS-PM 1*, and *EOS-CHEM 1* by one year, thus delaying the launch of second-wave spacecraft. With the report in hand, Kennel convened a second group of scientists who had no connections with MTPE/EOS and asked them to review the scope, balance, and risks associated with the rebaselined program. This assessment was completed the following month and was generally favorable to the EOS Payload Advisory Panel findings. It was believed that the new program would still be capable of meeting IPCC science objectives despite the constrained budget environment.[27]

As a result of the reviews carried out during the summer and fall, the rebaselined EOS program was integrated into the space agency's FY1996 budget request. In addition to the changes to EOS, this new budget also included NASA's new approach to earth system science. After the rescoping exercise had been completed, the program's allocation for science had fallen by 30 percent. This outcome was perhaps inevitable given the budget cuts that the program was facing as it ramped up the *EOS-AM 1* development. This modification, however, led Bob Harriss to rethink the focus of the space agency's earth science mission. He wanted to concentrate on developing a predictive capability for climate change, but the reality was that this capability was at least ten years away. Therefore, NASA would develop modest satellite data applications that would help maintain political support for MTPE. The feeling was that the agency could attract congressional support by focusing on applications that would assist in agricultural planning, predicting seasonable weather variations such as El Nino, responding to natural disasters, and establishing land-use and fishery policies. In many ways, this effort would simply build upon the efforts that had made Landsat and Nimbus so popular.[28]

Some saw this as a shrewd political decision, aimed at obtaining the support of legislators for whom the original program provided no immediate electoral benefits. One potential convert was Senator Conrad Burns (R-MO), who told *Science*, "The program will easily pay for itself in lives and property saved and improved water management." The scientific community, however, was critical because many believed the change in focus would lead to a loss of credibility, as MTPE had initially been presented as a global change program. OSTP Associate Director for Environmental Sciences Bob Watson said at the time that most scientists were disgusted with NASA's new MTPE science approach.[29] The strategy, however, seemed to pay immediate dividends. In September, the House-Senate conference committee agreed to increase the FY1995 EOS budget by $38.1 million above the president's request for a rescoped program.[30] During final budget negotiations between administration officials and the space agency a few months later, the OMB followed suit and matched that level of funding in the following year's presidential request. This was the first good news that the project had received in quite some time.

Just as NASA was struggling to secure EOS's future, other important developments within the federal government were impacting global change research. During the initial decades of the space age, the American space program had developed two autonomous weather satellite programs—one operated by NOAA and one operated by the DOD. A large number of studies had been conducted to explore options for merging the two programs, but they had not resulted in actual programmatic changes. In February 1993, Rep. George Brown had asked NOAA Administrator James Baker to commence another review to assess possible cost savings.[31] Around the same time, Senator James Exon (D-NE), chair of the Senate Subcommittee on Nuclear Deterrence, Arms Control and Defense Intelligence, had asked Commerce Secretary Ron Brown to review the two weather satellite programs to determine whether they could be combined.[32] The following month, President Clinton had initiated an ambitious review of the federal government to deliver on a campaign promise to "radically change the way government operates—to shift from top down bureaucracy to entrepreneurial government."[33] Gore had been given the task of creating a government that worked better and cost less. To accomplish this objective, he had assembled a team of career government executives and outside management consultants and embarked on the National Performance Review. In September, White House and congressional efforts came together when Gore's review team proposed combining the DOD and NOAA polar-orbiting weather satellite programs.

This change had important repercussions for NASA. Due to Gore's deep familiarity with MTPE, he proposed that the space agency play a key role in the new program:

> The Earth Observing System (EOS) [is] a series of six different satellites measuring various parameters critical to understanding global change. One of these satellites is called EOS-PM. The climate monitoring instruments on EOS-PM are basically more modern versions of the mete-

> orological instruments currently flying on the NOAA weather satellites. The EOS-PM climate research satellite is being designed with the idea that many of the instruments can be used by NOAA within its Polar Orbiting Environmental Satellite [POES] program. This continues a historical NOAA-NASA relationship wherein NASA develops new technology and demonstrates prototype hardware, and NOAA buys identical units for continued operational support. However, current plans involve flying EOS-PM for 15 years, during which time POES will also have operational satellites. Over most of this period, both programs would be flying duplicate instruments. The nation would be more efficiently served if NASA would develop and fly the prototypes once and then transfer the systems to NOAA's operational program for future flights.

On 5 May 1994, the Clinton White House issued Presidential Decision Directive NSTC-2, "Convergence of U.S. Polar-Orbiting Operational Environmental Satellite Systems." This document directed NOAA and the DOD to establish a single operational meteorological system that would eventually be called the National Polar-Orbiting Operational Environmental Satellite System. The directive established an Integrated Program Office that would coordinate activities between NASA, NOAA, and the DOD. NASA would have "lead agency responsibility to support the Integrated Program Office in facilitating the development and insertion of new cost effective technologies that enhance the ability of the converged system to meet operational requirements." In this vein, the space agency was expected to make EOS-PM instruments available for use on operational missions as soon as practicable. The directive provided that NOAA would be the lead agency for overall operations, while the DOD would play a lead role in systems acquisition.[34]

That spring, Al Gore determined that it was time to disband the Committee on Earth and Environmental Sciences. He remembered the struggles that had occurred between the Bush administration and congressional Democrats during deliberations over passage of the Global Climate Change Research Act. He was also unhappy with the USGCRP's failure to heed the NRC recommendation, from several years earlier, that steps be taken to improve the means by which federal agencies turned research results into usable information for policymakers. As part of the reinventing government program, therefore, Gore recommended a broad reorganization of the White House science policymaking process. This reorganization included a new coordinating committee, the Committee on Environment and Natural Resources, that was designed to be what its predecessor had not been—a body to coordinate policy alternatives for dealing with and responding to global change. Instead, the creation of this new committee ultimately led to the effective breakup of any interagency program for global change research. Among the problems that emerged was that the Committee on Environment and Natural Resources would have three cochairs rather than its forerunner's single chair, creating confusion regarding who was actually in charge. In addition, the committee was not tasked with budget coordination, which greatly reduced its ability to promote new policy alternatives. Finally, the policymaking effectiveness of the organization was badly fragmented when Cochair Bob Watson established more than a dozen subcommittees, each representing a different constituency. As Henry Lambright wrote a few years later, the result was that "the interagency coalition behind the Global Change Research Program began to unravel." Despite the proenvironmental stance of the Clinton administration, "from the committee's perspective, there was a sense of diminishing support."[35]

## Republican Takeover and More EOS Studies

In November, Republicans swept into power in both chambers of Congress. In broad terms, this change was a mixed bag for NASA. While GOP leaders quickly made it clear that they intended to continue government support for basic scientific research, they also suggested that applied research and development were best left to the private sector. Robert Walker (R-PA), the new chair of the House Science and Technology Committee, Jerry Lewis (R-CA), the new chair of the House appropriations subcommittee for NASA, and Christopher Bond (R-MO), the new chair of the Senate appropriations subcommittee for NASA, were all supporters of the space station. The problem for the space agency was that Walker, who would emerge as the most important new actor in the policy area, was not a fan of MTPE. In particular, he did not like EOS. He believed it was too expensive for a program dedicated to the study of global warming. On 15 December, even before assuming his new position, Walker expressed his uncertainty about the future of EOS: "I don't know if it needs redesigning but I do believe it needs to be looked at to make certain we are moving in the right direction." He suggested that he wanted to see more private industry involvement in MTPE: "I don't think everything we do in space has to come through the front door of NASA. I believe there are lots of things that can be done outside of NASA, and NASA can be reconfigured to fit that system." He would later tell reporter Andrew Lawler, "I thought what we were funding was a macro-environmental program. [Global warming] is too narrow for the sort of money we're investing." These were troubling signs for the future of MTPE/EOS.[36]

In January 1995, a White House directive ordered NASA to reduce its total budgets from FY1997 to FY2000 by 3 percent, 5 percent, 7 percent, and 9 percent, respectively. These decreases were proposed to enable a middle-class tax cut, part of a general strategy employed by the Clinton administration in an effort to be proactive as the Republicans took control of Congress. The reductions came as a shock to Administrator Goldin and his leadership team. "Goldin, who had become Clinton and Gore's 'poster boy' for agency reinvention and budget cutting, was being rewarded for his cuts with more cuts." What made these spending targets an even greater challenge was a deal between the White House and Goldin that exempted the international space station, which had become a crucial foreign policy tool when the Russian Space Agency had joined the program two years earlier. The total spending cut over four years would be $4 billion, but Goldin decided that he did not want to impose any further cuts to programs. Instead, he targeted NASA civil servants, support contractors, and facilities in an effort to change "how NASA did its business, not what it did."[37]

The previous spring, President Clinton had signed a presidential directive that tasked the National Science and Technology Council with reviewing all national laboratories operated by the DOD, Department of Energy (DOE), and NASA to ensure that their missions reflected administration priorities and evolving national needs. The NASA Advisory Council conducted this review for the space agency. When the council began its review in June 1994, it set out to assess activities at all ten field centers, examining the full range of agency research and development efforts (human spaceflight, space science, earth science, aeronautics, and advanced technology). It also examined the space agency's management practices, facilities usage, strategic planning, mission selection, and quality of science programs. The goal was to make recommendations for facilities closures (including field center closings), program cancellations, new initiatives, management changes, and processes for priority setting.[38]

On 28 February 1995, the NASA Advisory Council review panel submitted its report to Goldin. The committee made more than four dozen recommendations, including ten that impacted MTPE. These included several proposals designed to improve internal and external coordination between the NASA Office of Space Science and other USGCRP agencies. The report noted, "While there is excellent science being pursued within MTPE, there is a lack of definition of scientific milestones and need dates that will provide the national policy process with the necessary information to make decisions in a *timely* manner." The report also noted that there was "little evidence of advanced technology development for and infusion into MTPE, which should be an ideal candidate for technology development because of the program's long-term nature (15 or more years) and multiple flight opportunities." The panel believed that innovation was critical to the *faster, better, cheaper* philosophy, which required that long-term advanced technology planning imbue MTPE/EOS.[39]

To some degree, this last recommendation was contrary to what the EOS blue team had proposed a few years earlier. At that time, it had been determined that it was necessary to limit the complexity of the second series of EOS payloads. This was to be achieved by essentially freezing their design and taking advantage of successfully demonstrated heritage hardware. With the publication of the NASA Advisory Council review, new voices were suggesting that although the restructuring and rescoping processes had trimmed budgets for the initial series of platforms, the second and third series might be pricier than necessary. This argument centered on the costs associated with maintaining the contractors and suppliers of the spacecraft buses, the electronics, and the instruments at a steady-state level for more than eighteen years. Kennel and Goldin both came to believe that those costs were too high to be acceptable in the existing political and fiscal environment. The former was also skeptical about how practical it was to fly three identical sets of instruments, as he believed that no two instruments ever built were truly identical. For these reasons, the two decided to examine ways to define EOS by the science it conducted rather than by its hardware configuration.[40]

Around this time, NOAA proposed moving its satellite operations center from Silver Spring, Maryland, onto the campus of the Goddard in Greenbelt, Maryland. The state's congressional delegation made it clear to NOAA that it would block any such move, as it wanted the NOAA presence to be part of an economically revitalized Silver Spring. Considering that an integrated program office for the National Polar-Orbiting Operational Environmental Satellite System had already been created, however, Goldin and Kennel were inspired to consider additional opportunities for collaboration between the two agencies before the launch of the first satellite in the series in 2004. A former White House official remembers the thinking at the time:

> Dan Goldin made a case that in order for him to go after the next generation of technological breakthroughs in Earth science instrumentation, he needed to free up funds. But unfortunately, the science community was increasingly viewing NASA as responsible for providing long-term observations with the same instruments. Hence, this meant that NASA was becoming increasingly an "operational" agency responsible for maintaining legacy systems. Goldin didn't want that, and his comments resonated with OMB and OSTP. He argued NASA should focus on [research and technology], and NOAA [should] focus on long-term operations. Hence, he wanted to transfer the future EOS platforms. . . . We tried to use the National Polar-Orbiting Operational Environmental Satellite System effort as a model for how DOD/NOAA/NASA could work together.[41]

In March, Kennel initiated four parallel evolution studies to investigate how MTPE/EOS could be changed to take into account the fiscal situation while not losing focus on high-quality science or technology development. The initial study was a review and update of the EOS science strategy. Kennel wanted updated science objectives (including the identification of the highest-priority science areas), an updated data acquisition timeline, and a plan for integrating the science of existing operational satellites (e.g., the UARS and ERBE satellite) with the upcoming EOS series. An EOS reshaping study would follow, established to

> examine innovative means of pursuing the second and third flights of EOS measurements, as well as examining means by which technology could be infused into other MTPE satellites and into the national satellite weather program. This study will examine ways of reducing size, power, and processing requirements in making the measurements, while maintaining data continuity and integrity. It will also examine the

role of the "common spacecraft," whose use as a common bus had been planned for both the first and second phase of EOS.

Kennel wanted to identify ways to save money by incorporating new technology advances (including EOSDIS, algorithm development, and spaceflight hardware). He also formed a NASA-NOAA Alignment Study composed of three teams (science, flight, and data systems) with personnel from both agencies tasked with investigating an observational and programmatic strategy for EOS follow-up missions. Three more teams were established to review the NASA-NOAA alignment from a perspective broader than that of MTPE/EOS. Finally, Kennel created a commercial dimensions study to determine whether MTPE could utilize and/or stimulate commercial involvement in data distribution through data purchasing agreements, government use of commercial assets, or commercial use of government assets. This study was intended to determine whether it was possible to drive down the costs of acquiring science data by exploiting industry technologies and information systems and by working with the commercial sector to develop new technologies that were mutually beneficial to government and business. Each study group was to provide preliminary results to an EOS Investigators Working Group meeting scheduled for June 1995.[42]

As these internal studies were being initiated, a series of external events was unfolding that placed the future of MTPE/EOS in jeopardy once again. On 6 April, Bob Walker sent separate letters to National Research Council President Bruce Alperts and Administrator Goldin. The chairman was concerned with the narrow programmatic focus on global warming and did not like the "down-to-earth" science approach advocated by Robert Harriss. In the letter to Alpert, Walker requested an NRC review of the program and asked Alperts to conduct "an assessment of the current observational strategy for the USGCRP, with particular attention to NASA's Earth Observing System." He wanted the review to focus specifically on "adjustments that might be necessary to respond to: (i) recent improvements in scientific understanding, (ii) the needs of the general user community, and (iii) the requirement for EOSDIS to develop into an operational system routinely accessed by various users versus the need to rapidly incorporate new and emerging technologies." The Goldin letter was more pointed. Walker asked, "Is the science that is being conducted as part of the MTPE/EOS program fully justified on strictly scientific terms? Is it peer-reviewed?" Kennel recalled later that this issue of peer review was a constant source of inquiry from Walker, who was not pleased with the fanfare that NASA research results received when released publicly: "He wants to know, is it peer reviewed? So the warning was unmistakable: make sure that you don't go out on a limb on any of these issues that you talk about in published data. And the other issue was: don't go out making a big deal about it; just publish it." Walker wanted both the NRC external review and NASA self-assessment completed by September, which would ensure that the recommendations would be available to influence congressional action on the FY1996 budget.[43]

The following month, Walker took advantage of his position on the House Budget Committee to exert his influence again. At his urging, the committee passed a budget resolution establishing federal spending targets aimed at reducing the deficit and balancing the budget from 1996 to 2002. The resolution targeted projects at NASA, the NSF, and the DOE. It specifically said that "applied research can and should be market driven and conducted by the private sector." The accompanying report noted that the committee advocated "policies that encourage faster private technology development as risk becomes better understood and more controllable." Utilizing this approach would lead to a $2.7 billion reduction in the MTPE budget, part of a larger $5 billion cut recommended by the committee. Walker said later, "My effort was aimed at assuring that we got new generations of satellites that I thought would be less expensive both to launch and to operate and yet would be more robust."[44]

The suggested cuts came as a surprise to NASA leaders because Walker had not proposed them through the regular appropriation hearings process earlier in the year. At hearings before the House Appropriations Committee, NASA had faced very few questions about EOS. The lone exception was a handful of inquiries from Representative Rodney Frelinghuysen (R-NJ) about the pace of EOSDIS development. Additionally, there had been no questions for the record following the hearings that foreshadowed any action on the EOS budget.[45] Likewise, hearings held during the same period before the House Subcommittee on Space and Aeronautics had revealed no threats to MTPE/EOS. In fact, despite his being the chair of the authorization committee, Walker had not even been present to discuss his concerns directly with NASA leaders or ask questions about the impact of additional budget cuts.[46]

Both Goldin and Kennel condemned the budget resolution, arguing that it would effectively kill MTPE because it would require canceling the second and third spacecraft series. Goldin told the *Washington Post's* Kathy Sawyer that the reductions would force the American people "to decide whether they want to have a space program" because if the proposed cuts were passed, "the next step [would be] to shut it down." He wrote in a *Space News* op-ed that this resolution would "dismantle the national approach to the U.S. global change research priorities . . . and, destroy [EOS's] basic feature—comprehensiveness. [More importantly,] it would condemn American scientists to pursuing an approach to environmental research that is more than a decade out of date."[47]

On 9 June, Walker received more ammunition for his fight against MTPE/EOS. A General Accounting Office (GAO) study that he had requested earlier was released. Entitled *NASA's Earth Observing System: Estimated Funding Requirements*, the report estimated that the total cost of EOS from 1991 through 2022 (when the final EOS satellite series would be deactivated) would be $33 billion. It echoed the NASA Advisory Council's Federal Laboratory Review report in noting that the space agency did not seem to have a plan in place to infuse EOS with new technology to potentially save costs. The new report did note, however, that several ongoing internal studies might provide recommendations for reducing future allocations. On 13 June, Kennel testified before the Senate Committee on Commerce, Science, and Transportation that NASA was already ahead of the GAO report's conclusions. This did not stop the House appropriations subcommittee from taking action. On 21 July, it recommended a $338.6 million cut to the MTPE FY1996 budget. The subcommittee provided no language to explain the cut. Around that time, Kennel told *Science's* Andrew Lawler, "If this is a down payment on the $2.7 billion cut, it basically stops EOS in its tracks after the first satellite is launched in 1998." Republicans who favored the cuts called this response a "Chicken Little" overreaction.[48]

On 19 July, the NRC's Committee on Global Change Research began meeting in La Jolla, California. The committee was chaired by Berrien Moore, Director of the Institute for the Study of Earth, Oceans and Space at the University of New Hampshire. Over the course of the subsequent ten days, the regular committee set about its task with the assistance of several ex officio members, including former NASA Administrator Robert Frosch, Francis Bretherton, and Edward Frieman. Charles Kennel flew across the country to present the results of the MTPE evolution studies, describing what he called a reshaped EOS. The assessments had identified nearly $70 million in annual savings that could be achieved by reducing NASA's on-site support contractor workforce, decreasing staff levels at NASA headquarters by nearly 40 percent (accompanied by the movement of program management responsibilities to the field centers), and sanctioning expected cuts to the total NASA civil service workforce. As a result, Kennel had identified $325 million in cuts from 1997 to 2000.[49]

The reshaped EOS would focus its science mission on collecting data to improve studies of atmospheric ozone, seasonal to interannual climate change, long-term climate variability, and land-cover change and global productivity. These were the scientific elements that NASA believed it was responsible for as part of the USGCRP. To inform these general areas, the space agency would have to maintain a set of twenty-four measurements. These data would be acquired by the first series of EOS satellites (EOS-AM, EOS-PM, and EOS-CHEM), *Landsat 7*, two single-instrument altimetry missions (requiring the deconstruction of *EOS-ALT* into two smaller satellites, one an international project), and the utilization of several international flights of opportunity for additional instruments. The study teams had determined that the first series of EOS satellites was already state-of-the-art and too far along their development paths to be changed. The agency agreed, however, that the second series of spacecraft had to be changed.[50]

For follow-up missions, NASA proposed a dual path. Along one path, the EOS-PM series would be a test bed for the National Polar-Orbiting Operational Environmental Satellite System. If the system proved successful, next-generation instruments would fly aboard platforms in this constellation. This would remove the burden for NASA of developing further EOS-PM satellites and would also renew a relationship between NASA and NOAA that had existed during the early space age, when the former funded initial innovations before transitioning proven capabilities for use in operational weather satellites to the latter. It was expected that the two agencies would also seek to collaborate on EOSDIS, saving additional funds by leveraging NOAA information technology capabilities, using common locations for ground stations, consolidating the information service infrastructure, and increasing automation to reduce staffing needs. Although the earlier NASA-NOAA relationship had proved to be unsuccessful, it was believed that this failure would not repeat itself because the Clinton administration had provided the agencies with a new mandate when it created an integrated program office for them. NOAA would not be required to accept NASA-developed technology if it would not be suitable for operational use. NASA could ensure that its research was on the right track, however, through regular interagency communications. Furthermore, the Clinton administration's reduction in total funding for earth observations provided an incentive for an effective partnership. NASA and NOAA would be forced to better manage the technology infusion process to identify cost savings and ensure data continuity. Most importantly, the leadership teams of both agencies had seemingly embraced this opportunity.[51]

Along the second path, the NASA team assumed that EOS-AM and EOS-CHEM instruments would fly on second-generation satellites that would be smaller and cheaper to develop. This could be accomplished either by splitting the instruments onto separate platforms or by utilizing second-generation instruments that were also expected to be smaller and cheaper. The space agency would create two avenues for advanced technology development for the earth sciences. First, NASA would locate the newly created Earth System Science Pathfinder Program within MTPE. These missions would be NASA-industry partnerships that enjoyed limited government oversight to encourage the rapid development of cheaper technologies. The total government mission cost would be less than $120 million. These more affordable technologies would form the foundation of second-generation EOS missions. Earth System Science Pathfinder missions would represent a genuinely new way for NASA to do business with the aerospace industry. While NASA's involvement with its contractors had always involved a significant amount of oversight, this new approach was formulated around the idea that providing more flexibility would produce cost savings and better spacecraft. Second, the space agency would channel funding to the New Millennium Program, a technology development program within the OSS created specifically to fund instrument technology developments. The hope was that this program would provide MTPE with smaller and cheaper instruments. Additionally, there would be a focus on improved measurement capabilities, simplified spacecraft architectures, and straightforward operations and data management requirements. In many ways, this was a return to the principles of NASA's earliest remote sensing programs in the 1960s. At that time, the aeronautics program served as the space agency's development program for new sensors, and new technologies were tested at high altitudes well in advance of flying in space. Because both technology and requirements had evolved since the 1960s, the goals of this new development program were somewhat different. Regardless, the principle was the same: test it first.[52]

In aggregate, this collection of changes was revolutionary. Essentially, NASA had abandoned the original EOS concept. As originally conceived, the program was designed to provide platforms capable of answering scores of questions posed by earth scientists. To do so, large EOS platforms would have guaranteed continuity in measurements by committing to a predetermined satellite configuration that would fly for a minimum of fifteen years. The reshaping exercise fundamentally changed the program. NASA stripped down the science mission to twenty-four core measurements. The second and third EOS series (copies of *EOS-AM 1*, *EOS-PM 1*, and *EOS-CHEM 1*) were also eliminated. To manage the transition to Earth System Science Pathfinder and New Millennium Program technologies, a MTPE biennial review process would be established. Each review would involve three specific tasks: an examination of all planned American, international, and commercial programs; matching those programs with the twenty-four measurement requirements; and revising mission profiles and acquisition strategies to meet MTPE science objectives. NASA's leaders hoped this formalized and proactive process would enable the space agency to preempt future congressionally directed reviews.[53]

Dan Goldin and Charles Kennel had essentially transformed an expensive hardware development plan into a more affordable program. After 2000, the two wanted MTPE spending to be capped at approximately $1 billion annually while maintaining the requirement for twenty-four measurements. They believed the Earth System Science Pathfinder and New Millennium Program investments would produce instruments that would enable NASA to accomplish this task. With this spending cap in place, the space agency would reduce its total costs through 2022, which the GAO had identified as the end of the EOS program, by at least 30 percent. Goldin later said he believed the new approach would make NASA earth science much more responsive to the changing needs of its scientific constituency:

> The program needed further restructuring because it was still much too unwieldy. . . . It didn't have an ability to exert more time responsive science, and allow the flexibility to the Earth science community to really be able to, in a rapid fashion, get the data that they needed. The dilemma with that was if you just launched those spacecraft on five year sensors, and given how long it would take to develop those platforms over a fifteen year period, you wouldn't have time to get data back and restructure the payloads for each series of spacecraft. So what we set out to do was see how we could decouple the payloads, so instead of having multiple payloads on a spacecraft . . . what we wanted to do is have single or double payload spacecraft. . . . That's where faster, better, cheaper came from.[54]

On 12 September, the NRC's Committee on Global Change Research released an initial report that was overwhelmingly positive regarding NASA's new plans for MTPE/EOS. The study team sounded a strong note of support for the first series of EOS satellites:

> The present review has concluded that structural change to the near-term EOS missions beyond the limits achieved in the 1995 reshaping exercise would cause severe program dislocations. Further budgetary reductions or imposed constraints on technical options could mean the elimination of key sensors, slips in schedule, loss of data continuity, and the elimination of advanced technology development that could enhance future research and lower costs. Our review has concluded that a shift to smaller platforms for the first group of instruments would be premature, since it could eliminate key instruments.

Furthermore, the committee endorsed the reshaped plans for more advanced technology and smaller spacecraft for later missions. The report supported NASA's plans for "down-to-earth" applications using EOS satellite data, stating, "The capabilities of these [satellite] systems will contribute to practical applications such as natural hazards mitigation, water resources management, and food and fiber production, as well as advanced Earth Sciences." The group noted its agreement with NASA's adoption of the decentralized approach for EOSDIS recommended by the Zraket committee. There were still concerns, however, that not enough control over data products had been given to the scientists. As a result, the report recommended that the "EOSDIS management and operations concept should be refined to involve the broader user community more effectively."[55]

The report was favorably received in the Senate, where Bob Harriss had been actively communicating with the members interested in the possible applications enabled by EOS. The day after its release, Senate appropriators gave MTPE a significant boost by recommending only a $61 million budget cut. The budget cut was directed entirely at EOSDIS, keeping it at its FY1995 budget level until there was an opportunity to implement the NRC's recommendations for new management arrangements. Citing the NRC report as justification for leaving the program essentially alone, the committee wrote, "The National Academy of Sciences has recently reviewed the Earth Observing System Program and reaffirmed the program goal and overall approach of providing scientific understanding of the Earth as an integrated system." This victory did not spell an absolute victory for MTPE. President Clinton and Congress were engaged in a high-stakes budget showdown, which in November led to a government shutdown. In the meantime, Dan Goldin established a Reshaping Options Implementation Study to review the recommendations of the NRC and the evolution studies reports.[56]

## Stability at Long Last

During the budget showdown, Bob Walker mounted a last-gasp effort to implement his proposed $330 million MTPE reduction for FY1996. Unimpressed with the outcome of the NRC review, he told Andrew Lawler that it was "not the be-all end-all," and he still wanted to implement changes. On 5 December, the House-Senate conference subcommittee for VA-HUD and Independent Agencies adopted its appropriations bill. Although it cut $81 million out of the MTPE budget in FY1996, the legislation was vetoed by President Clinton for reasons that had nothing to do with the space program or climate change research. In January 1996, Congress began to compile an Omnibus Appropriations Bill in an effort to obtain some movement on the budget front. Walker used this as a chance to make "back-door attempts to cut the EOS budget." He approached House appropriators who were working on the omnibus bill and asked them to cut an additional $156 million from the MTPE budget and introduce language that would block NASA from signing a contract for the common spacecraft development for *EOS-PM 1* and *EOS-CHEM 1*. This move failed because House VA-HUD Appropriations Subcommittee

Chair Jerry Lewis (R-CA) objected. Lewis told Lawler, "This was not a joint effort. I know Bob's concern, and if he had his way, he would choose to eliminate funding." Lewis thought this was pointless because Senate appropriators opposed these budget cuts. As these discussions were being carried out within the appropriations subcommittee, Walker was also working on an authorization bill in the House Science Committee that would have forced the space agency to hold off on spending most of its FY1996 EOS appropriation until a congressionally appointed review panel assessed the scientific merits of MTPE. This measure failed as well because Senate authorizers would not agree to such a provision. In the end, no authorization bill was passed for FY1996. In April, seven months into the fiscal year, an Omnibus Appropriations Bill was finally agreed upon and signed by President Clinton. It cut the EOS budget by $91 million, necessitating a one-year delay in the launch date of *EOS-PM 1* and *EOS-CHEM 1*. These schedule delays, however, did not change the technical configuration of either satellite.[57]

In February, while budget negotiations were still ongoing, the Reshaping Options Implementation Study team completed its work and strongly endorsed the broad, strategic conclusions of the MTPE evolution studies. The group recommended a number of tactical, organizational, and management approaches to oversee technology development and infuse the user community more effectively into EOSDIS. While some follow-up action was required, it was clear that the basic approach outlined by Kennel and presented to the NRC team would be enacted. As the OMB was preparing to submit the FY1997 budget to Congress, it reached a delayed agreement with the space agency that would adopt the reshaped EOS plan and provide a $1.4 billion allocation. It was expected that funding between FY1996 and FY2000 would total just over $6.8 billion.[58]

During 1996, Bob Walker and House Republicans continued their efforts to drastically cut EOS. Each time, Senate Republicans restored the funding. By the end of the FY1997 appropriations process, although the House-Senate conference committee had cut $5 million in funding for an international MTPE initiative known as Global Learning and Observations to Benefit the Environment (GLOBE), it otherwise directed "no specific reduction to Mission to Planet Earth programs." Congress continued to support the earth probes missions, plans to use MTPE funds in the New Millennium Program, and plans for commercial partnerships through the Earth System Science Pathfinder Program. This was the first year since

1991 that the space agency was not forced to conduct a significant EOS "re-" effort or to negotiate significant out-year budget reductions with the White House. At the end of the year, Bob Walker retired from Congress, and Charles Kennel left NASA. Three years later, *EOS-AM 1* (renamed *Terra*) was successfully launched from Vandenberg Air Force Base into a sun-synchronous orbit around the earth and began returning data to terrestrial researchers.[59]

# Conclusion

As we discussed in the introduction, our objective for this study was not solely to provide a narrative history of the MTPE/EOS program. While the retelling of this fascinating story helps us understand how and why this critical undertaking came into being, we felt from the beginning that this treatment would be incomplete without also examining the policy process through a theoretical lens. In this conclusion, we return to the foundational models presented in the introduction in an attempt to better grasp the events portrayed in the narrative, beginning with a discussion of how the multiple streams and punctuated equilibrium models can help explain how and why MTPE/EOS reached the national agenda and was ultimately adopted. This will be followed by a lengthier consideration of how the life cycle of bureaus literature can provide us with a deeper understanding of how one of NASA's strong external clients was critical to the adoption and protection of a significant earth science mission. The final section of the chapter will consider the central lessons that emerge from our research into the history of the MTPE/EOS program.

With his multiple streams model, John Kingdon suggested that joining three process streams (problems, policies, and politics) is required to gain support for the adoption of any new initiative. He describes a "policy primeval soup" wherein problems and solutions float about, waiting for policy entrepreneurs to mold them into agenda items capable of gaining support within the political stream. Before seeking adoption, however, these policy entrepreneurs take time to soften up decisionmakers. As Frank Baumgartner and Bryan Jones assert, this task allows the policy entrepreneur to draw new attention to a problem and enlarge the policy community interested in finding solutions. All three view the softening-up period as critical to success and contend that it often takes many years and requires consistent dedication. This was certainly the case with MTPE/EOS, as shown in detail in the first three chapters of this book.[1]

In some ways, the softening-up period for MTPE/EOS dates all the way back to the establishment of weather satellite programs by NASA starting in the late 1950s. It became more defined with the battle for approval of *Landsat-1*, which introduced the notion that satellite remote sensing could play an important role in environmental monitoring. During the 1970s, the introduction of additional spacecraft and instruments designed for remote sensing began to influence the development of a global perspective of the earth's climate. Data from these projects was used, in particular, to augment land-based observations being made by earth scientists such as Charles Keeling. This led to the early establishment of the UARP and the development of the UARS—although both had quite small budgets compared with what was to come in a couple of decades. During the early 1980s, NASA began to take more aggressive steps to soften up the relevant policy communities. This began when the space agency tasked the NRC's Space Studies Board with reviewing mission priorities that might lead to a dramatically enlarged role in earth science research. It continued with ill-fated efforts, including the Goody and Friedman Reports, to explore pathways that would allow NASA to become more involved in global habitability studies.

By the mid-1980s, the space agency's softening-up campaign had gone off the tracks a bit as insiders attempted to attach climate research to the Space Station Program, an idea that was called into question after the *Challenger* accident. This tragedy resulted in NASA returning to square one to determine its strategic goals for the next couple of decades, which included convincing key policy actors that it had an important role to play within the global change research realm. The two Bretherton committee reports and the Eddy Report were part of the effort to stay on track during this period, specifically by providing road maps for space-based environmental monitoring. Both reports offered broad-based ideas for suites of spacecraft and instruments that would allow the space agency to be an important player in the earth sciences. Yet for much of the remainder of the decade, EOS was sucked into the vortex of space station planning. In 1987, the program was finally decoupled from the Space Shuttle Program when Administrator Fletcher made the decision to switch to expendable launch vehicles to place EOS satellites into orbit. In 1989, EOS was similarly separated from the Space Station Program because it was determined that it would be less costly to carry out as an independent project.

One could argue that the softening-up period for a policy problem ends when a policy window opens. However, policy windows are not generally binary unless there is some kind of earth-shattering focusing event, which happens very rarely. The best way to understand a policy window is to picture a typical sash window in a house, where the inhabitant can slide one of the panels up or down as dictated by the weather outside. Policy windows operate in a similar fashion, although policy actors have far less control over when and how much the panel moves. The window may begin to open with some change in either the problem or political stream but begin to fall shut again if no momentum is generated toward adopting a feasible solution. Starting in the mid-1980s, there were several occasions when a policy window began to open that might have led to the embrace of a more significant earth science mission for NASA.[2]

After the failure to unlock a window for an ambitious global habitability initiative, the first time that another window began to open was during the ozone debate in the mid-1980s. The revelation that 40 percent of atmospheric ozone had been lost and that a giant hole had opened above Antarctica was the first focusing event that began to shift the national mood with regard to support for a Big Science research project that would study global change. An overly cautious scientific community ultimately squandered this opportunity by failing to urge immediate action, but even as the window began to slowly close again, there was now a gap that future policy entrepreneurs could attempt to enlarge. The next focusing event that made this possible was the *Challenger* accident, which forced an all-encompassing review of national space policy—first by the National Commission on Space and ultimately by Sally Ride's strategic planning team. The Ride Report, for the first time, suggested that MTPE should be viewed as one of the central organizing options for the space agency in the coming decades. This pushed the policy window open even further, but not quite enough to join the streams and shove a fully funded initiative onto NASA's budget. While some small components of EOS progressed through the early procurement stages, the certainty needed to truly develop a comprehensive global change program was still lacking. The election of George H. W. Bush was the decisive change within the political stream that opened the policy window wide enough to gain support for MTPE/EOS. Environmental policy had been a central issue for Bush's campaign, and upon taking office, he made establishing the USGCRP a key presidential priority. The only question was whether NASA would play a role. It quickly became apparent that the new president would support the space agency's plans but the level of this support was not totally clear. After many months of wrangling with the OMB and the CES, Richard Truly was able to successfully secure funding for MTPE/EOS within the FY1991 budget submission to Congress. Although many years of bureaucratic and congressional skirmishes lay ahead, Truly had been able to leverage the many years of work that had been done within the policy primeval soup to push MPTE/EOS fully through the policy window and gain approval for one of the greatest Big Science programs in American history.

Baumgartner and Jones argue that the policy process experiences long periods of stability during which only incremental changes occur, which are punctuated by short periods of instability that result in a significant shift in the scope or direction of government activities. This pattern can clearly be seen in this book's historical narrative. Dating back to the 1950s, NASA's policy agenda had only three major components—human spaceflight, space science, and aeronautics. While incremental changes were made over time to slowly include some earth science research, these missions were comparatively small. During the mid-1980s, this dynamic began to change. Why did this happen? There were at least two reasons. First, the space agency's guiding mission had been drifting since it had successfully landed humans on the moon. Second, increasingly compelling land-based observations were introducing the idea of global climate change to the American public and political elites. This led many to believe that there was a need for better research to determine the severity of the problem and that space-based assets could be utilized to acquire needed data.

Baumgartner and Jones established the Comparative Agendas Project (CAP) to provide scholars with the resources necessary to longitudinally investigate trends in policymaking. Below, we will employ a mixture of policy image and venue indicators to better understand why MTPE/EOS was finally adopted in the late 1980s. This data also explains why it took so long to navigate congressional and bureaucratic storms before the overall mission budget achieved a relative level of stability. These results provide a surprisingly consistent picture regarding the potential for MTPE/EOS to reach the agenda and obtain support for ultimate approval.

Baumgartner and Jones elected to study media coverage to gauge trends in policy image. The primary information source that they utilized was the New York Times Index. CAP researchers have coded all of the articles written in a given year to set the context for the agenda process within a specified issue area. One concern was that looking at only one index would not fully capture the nature of public opinion regarding specific issues. Baumgartner and Jones found, however, that "when we compare levels and tone of coverage . . . it makes little difference which index one uses." Over 2,500 articles were coded with regard to climate change and global warming during the two decades that are the focus of this book's historical narrative. As can be seen in Figure 1, the amount of media attention given to climate change during the Reagan administration was extraordinarily limited. This began to change in 1988, as the number of articles nearly quadrupled at the same time that Governor Michael Dukakis and Vice President George H. W. Bush were campaigning for the presidency. In the following two years, coverage more than doubled again. Given that the issue was receiving so much coverage, it makes sense that President Bush made USGCRP and MTPE/EOS central initiatives within his overarching environmental agenda. As can be seen, however, as soon as this was accomplished, the reporting plummeted severely for a number of years. This reduced attention made it easier for Congress to stall implementation while at the same time hacking away at the original budget provided for MTPE/EOS. It was not until the amount of attention dedicated to the climate crisis began to increase again after President Bill Clinton's reelection that NASA was finally able to achieve budget stability for the initiative.

A somewhat similar pattern materializes from an examination of survey research, which, if available in a systematic form, can be used to observe changes in policy image. We have elected to utilize the Gallup Poll, specifically because it has for many decades asked respondents what they believe to be the most important problem facing the nation. This data reveals interesting trends in mass public opinion with regard to the environment. As can be seen in Figure 2, concerns about the environment never reached the same heights as economic anxieties during the period that we are investigating. In fact, such concerns barely register for much of this era. What is interesting, however, is that a relatively significant spike in environmental

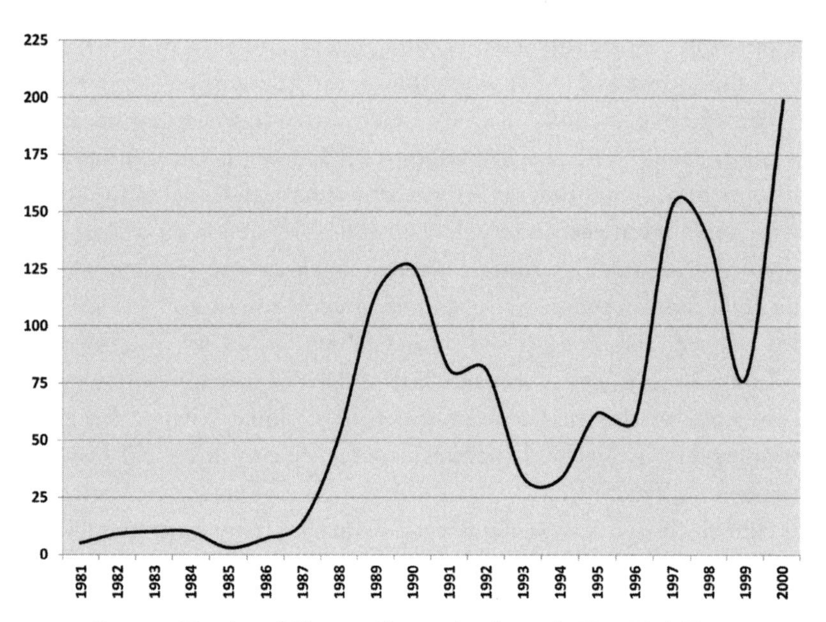


*Figure 1: Number of Climate Change Articles in the* New York Times

unease occurred around the time that MTPE/EOS was adopted. From 1985 to 1988, less than 1 percent of respondents believed the environment to be an important problem for the government to tackle. But in 1989, fully 4 percent of those surveyed considered it to be a critical issue. This number stayed above 2 percent for the next few years before settling at around 1 percent for the remainder of the decade. Once again, this suggests that at the time MTPE/EOS was gaining final approval, this type of governmental effort was deemed important by an increased, if still quite small, subset of the population.

While examining media coverage and surveys provides information related to policy image, other measures are needed to evaluate venue access. Venue access gauges the degree to which political elites are engaged with an issue. Two different indicators were selected for this book to monitor changes in venue access for environmental issues. First, we examined data from the Congressional Information Service Abstracts (CIS annual), a yearly compilation of all congressional hearings. CAP researchers have coded nearly 100,000 congressional hearings using a common policy content code to ensure compatibility over time. Second, we examined data from the annual State of the Union address to assess the level of presidential attention. CAP researchers have coded more than 22,000 quasi-state-

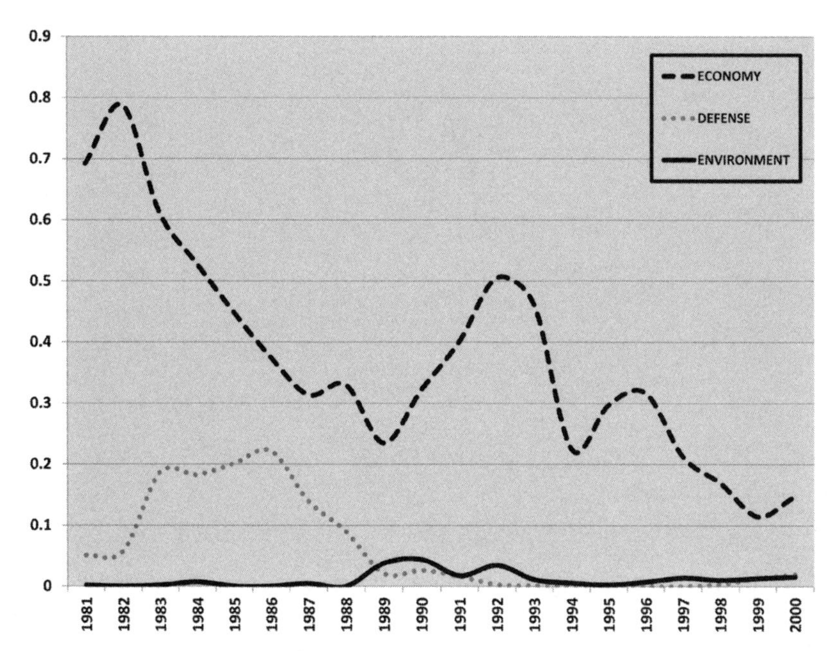


*Figure 2: Percentage of Americans Mentioning Issue as the Nation's Most Important Problem*

ments in these speeches, again using common policy content categories. As was the case for policy image, the venue access data reveals a political ecosystem that was ready to tackle the need for better climate research and to determine what NASA's role in this arena would be in the future.

During most of the 1980s, as can be seen in Figure 3, Congress convened on average about seventy-five hearings on environmental topics every year. This number was relatively modest because it was understood that President Reagan had no interest in working on environmental issues. This changed dramatically over the next six years as a result of fundamental changes within the political stream. First, Democrats retook the Senate and therefore had control of both chambers of Congress. Second, President Bush and President Clinton were willing partners in efforts to pass needed environmental legislation. Therefore, the average number of annual congressional hearings reached nearly 125 during this period. This meant that efforts to create the USGCRP and MTPE/EOS were now possible, and key policy entrepreneurs rushed to push them toward adoption while this window of opportunity was open. After the Republicans recaptured both chambers of Congress in 1994, the number of hearings fell by more than

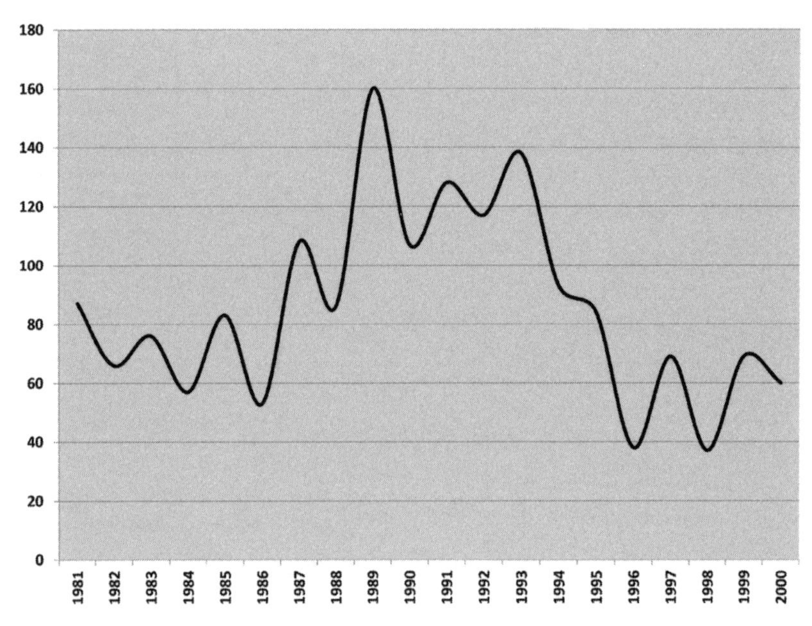


*Figure 3: Number of Congressional Hearings Related to the Environment*

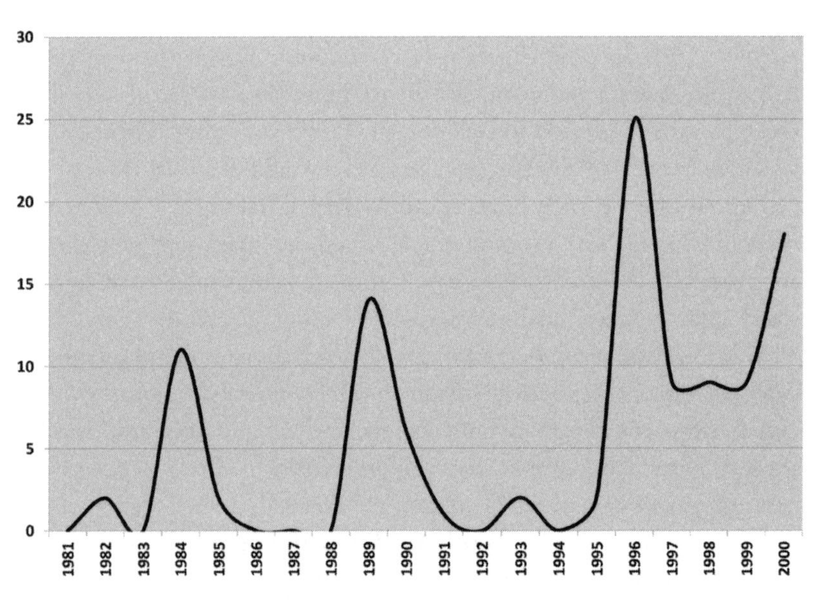


*Figure 4: Number of Environmental Statements in the State of the Union*

50 percent. Not surprisingly, therefore, this was a period when MTPE/EOS struggled for survival while attempting to achieve some measure of budgetary stability, which came about only after Clinton was reelected.

The number of statements in the State of the Union address related to the environment presents an even more interesting story. For the majority of these presidential addresses during the period of study, as can be seen in Figure 4, there were at most a few mentions of issues related to the environment (in seven of twenty, there was not a single reference). Therefore, when did the environment show up in these speeches? Three of the four sizable spikes in the number of statements were seen during years when a presidential election was being contested, suggesting that the issue became "important" only when there was an attempt being made to court voters dedicated to environmental protection. It is interesting that when he sought reelection, George H. W. Bush did not run on his impressive environmental record and lost. Contrast this with Bill Clinton, who did run on his far more lackluster achievements and won. To his credit, after being reelected partially on the basis of his promises with regard to the environment, Clinton did become a more consistent supporter of environmental initiatives. This can be seen in the steadier number of mentions in his State of the Union addresses, which likely had a positive impact on the stabilization of the MTPE/EOS initiative toward the end of the 1990s. What is most striking in the State of the Union data, however, is the dramatic jump in the number of environmental statements in George H. W. Bush's 1989 address, which is a clear outlier. While his predecessor had commented on environmental issues a grand total of fifteen times during his eight State of the Union addresses (eleven while seeking reelection), Bush discussed this topic fourteen times in his first speech before a joint session of Congress. This suggested that there was unusual access to the presidential agenda for environmental problems and solutions. As a result, Richard Truly and other policy entrepreneurs were able to gain approval for both the USGCRP and MTPE/EOS. The upshot was the addition of a multibillion-dollar earth science initiative to NASA's budget, an achievement that the space agency had been working to accomplish for more than a decade. All four of the policy image and venue access indicators explored above suggest that this was the very definition of taking advantage of a brief punctuation in the otherwise stable policy process within the space and environmental policy communities. The outcome was the adoption of a Big Science program that has been fundamentally important in providing society with a better understanding of the contours of earth science and the climate crisis that we all face.

## NASA as an Aging Client Bureaucracy

In his classic work on bureaucracy, Anthony Downs suggests that when a new government agency is established, its leaders must seek to manipulate the surge in enthusiasm that brought about its creation. He argues that this is necessary because such an organization is at its most "vulnerable to annihilation by its enemies immediately before it attains its initial survival threshold." To continue to exist, a bureaucracy must expand its sphere of influence and at the same time establish its autonomy. It is best suited to do so, Downs contends, when it is founded. If its rivals are successful at halting either of these efforts, the agency may not survive. James Wilson suggests, however, that the agency must be strategic in choosing where to seek expansion opportunities. In so doing, he reasons that the bureau should remain faithful to its mission, as expressed by the legislation upon which it was formed, when it considers taking on new tasks that are not already being performed by other organizations.[3]

At the beginning of the space age, there was no bureau clearly responsible for developing civilian earth observation satellites. With the passage of the National Aeronautics and Space Act of 1958, this task was given to NASA. Given that all nonmilitary satellite projects were transferred to the new space agency, it could reasonably define atmospheric research as part of its mission. This new responsibility, however, did encroach upon activities that were being conducted by other organizations, particularly the Weather Bureau. As long as NASA was interested only in satellite development, there would be no overlap with the Weather Bureau. However, this was not the limit of the space agency's ambitions. With Nimbus, it had conceived an advanced satellite intended to provide a variety of necessary meteorological data. NASA's plan to build an operational system using this technology impinged on the bureaucratic turf of the Weather Bureau, which had been given responsibility for coordinating meteorological activities seven decades earlier. This move also trespassed upon military programs occupying the same space. Regardless, NASA was able to wage an insurgency against these more established adversaries to gain congressional and presidential support for an important role in the establishment of early weather satellite systems via the Nimbus series.[4]

As Wilson argues, to secure its autonomy, an agency must fight other organizations to secure its sphere of influence. It should also be wary, he wrote, of cooperative programs. These can be used to slowly erode an organization's turf. The decision to rely on Nimbus technology left the Weather Bureau dependent on the new space agency, a position that no established bureaucracy would find acceptable. Thus, the Weather Bureau came out swinging with a new proposal to pursue an operational satellite program of its own using TIROS technology. The Weather Bureau was able to make such a move because it had support from the DOD, as the latter had agreed to provide the former with needed launch services. This offensive had the desired effect. NASA was trying to develop supremacy in civilian launch services and had no desire to face off with such a formidable competitor as DOD. At the same time, the space agency's leaders understood that it did not make much sense to push forward a satellite development program that no user wanted. Thus, despite the political and public support that it enjoyed, NASA found itself in a position where its only viable option was to establish a cooperative relationship with the Weather Bureau. The ultimate compromise involved the space agency developing and launching weather satellites, while the Weather Bureau would retain day-to-day operational control. This cooperative autonomy was the closest that each organization could come to defining a self-directed existence given the overlap in their bureaucratic territories.[5]

For a variety of reasons, this compromise actually worked for NASA. First, it gave the space agency the opportunity to carry out the Nimbus program unencumbered by Weather Bureau mission requirements. This allowed it to pursue a more innovative suite of observation technologies and sensors, establishing it as the most important agency for atmospheric research within the federal government. Second, the arrangement provided the less established organization with needed experience in dealing with a user that would eventually utilize the resulting operational satellites. Although Wilson suggests that agencies should avoid cooperative ventures, NASA was forced into this venture, and it ultimately helped the space agency cross its bureaucratic survival threshold within this mission area. The important role that the organization gained in atmospheric research would serve it well in the decades to come.

### *NASA as an Aging Bureaucracy*

Downsian bureaucratic theory suggests that after their creation and rapid initial growth, all administrative organizations begin a marked decline in a number of areas. Downs argues that after the early growth phase, competition for resources with other agencies and the difficulties inherent in managing an increasingly large organization lead to degeneration. This is seen in the slower recruitment of top talent, an aging workforce, the addition of new layers of senior management, and the adoption of a generally conservative culture. Downs refers to this last point as the "Law of Increasing Conservatism," which is the result of changing personalities within a bureau. When it first starts out, an agency is filled with aggressive advocates, or "climbers," who seek to innovate as a way to spread the influence of the bureaucracy. As it ages, however, an organization's leadership looks to protect its position and become "conservers." Downs writes, "All bureaus tend to become conservative as they grow older. . . . As a result, the entire bureau will shift toward greater conserver dominance, thereby reducing its ability to innovate." NASA's birth and rushed expansion happened over a period of less than a decade—roughly from 1958 to 1965 (the latter date being the high point of funding for Project Apollo). After this initial burst of activity, the agency experienced a relatively steady decline for more than a half century.[6]

To some degree, Burt Edelson was trying to partially reverse this trend when he added polar-orbiting remote sensing satellites to the agency's decision agenda during the early 1980s. The problem was that even as some within the organization were trying to gain support for an exciting new mission in global change research, they had little support from President Reagan. During a Senate hearing in June 1986, Senator John H. Chafee (R-RI) characterized the mood:

> It was the scientists yesterday who sounded the alarm, and it was the politicians, or the government witnesses, who put the damper on it. We don't have all the perfect evidence. . . . Nonetheless, I think we have got to face up to it. We can't wait for every shred of evidence to come in and be absolutely perfect; I think we ought to start to try and do something about it, and certainly, to increase the public's awareness of the problem.

Senator George Mitchell (D-ME) was even more blunt: "We are again faced with the dismaying prospect of an administration policy that everything must be known before anything can be done. It is a familiar pattern, one which the members of this committee have repeatedly been exposed to in many areas dealing with environmental policy." This idea of the scientists versus the bureaucrats was strongly echoed in the press coverage in the *New York Times* and *Washington Post* that followed the hearings.[7]

As this chorus of criticism increased, President Reagan was essentially forced to respond. A former White House aide recalled that "this is not the way Reagan asked the question, but the question was basically: 'Is there anything to this climate change issue, and if there is, what am I, as President of the United States, supposed to do about it?" Ultimately, he directed his Domestic Policy Council to form a working group on climate change led by Anthony Calio. This was the first time that significant presidential attention was paid to the issue, and it led to the creation of the CES. In reality, however, this was an incredibly weak first step, and its weakness was a reflection of the conservative leanings of both President Reagan and Science Advisor William Graham. The latter supported the establishment of the CES because he hoped that an "active global change committee would lead to a . . . more cautious and conservative . . . intellectual atmosphere throughout the government." Collectively, the Reagan administration's lackluster response to the growing climate crisis significantly delayed the adoption of MTPE/EOS. At the same time, it seemed to confirm concerns that the space agency was an aging agency incapable of taking leadership within innovative new research areas.[8]

In 1990, the Augustine committee acknowledged NASA's status as an agency in decline when it wrote in its highly influential report,

> There is the matter of institutional aging and the concern that NASA has not been sufficiently responsive to valid criticism and the need for change. . . . The civil space program has gradually become afflicted with some of the same ailments that are found in many other large, mature institutions which have no direct and immediate competition to stimulate change. . . . NASA no longer operates under the relatively more flexible policies, regulations, and legislative environment that characterized its early years.

The commission noted that projects were increasingly fashioned to protect the existing workforce rather than having the workforce change as required by new mission needs. It concluded, "Bigness is not itself goodness . . . the natural tendency of most engineering pursuits seems to be toward bigness [but] specific guards must be established against unjustified growth." The commission's findings had a significant impact on the Bush administration, which ultimately led to the replacement of Richard Truly with Dan Goldin as the space agency's administrator. This change had important impacts on earth science research.[9]

Shortly after the Augustine committee released its report, Howard McCurdy published a detailed study of the evolution of NASA's organizational culture. His research revealed that the space agency demonstrated many of the characteristics of a declining bureaucracy, including an aging workforce that showed the tendency to lose climbers and maintain conservers. McCurdy argued that these conservers became "lifers," and as a result, the agency had little to no turnover through the 1980s. During this period, NASA's approach to program management became "more bureaucratic in the pejorative sense of excessive paperwork and a preoccupation with official procedures." One piece of evidence that McCurdy used to support this argument was the fact that the number of people classified as professional administrators increased from 5.5 percent in the 1960s to 18 percent in 1988. As this shift was happening, NASA's technical culture also changed as its leadership became increasingly cautious, less tolerant of failure, and predominantly concerned with bureaucratic survival. Even more disturbing, McCurdy found that "NASA engineers and scientists of the second generation retained many beliefs from the first generation even as they lost the ability to practice them. . . . A culture that permits such a gap between belief and practice cannot persist for long."[10]

Downs argues that the factors that cause organizational decline are often "rooted in exogenous factors." The Augustine committee contended that the most important external dynamic that impacted the post-Apollo space agency was the lack of a clear national consensus regarding the appropriate objectives for the civil space program. NASA's budget had begun to decline several years before humans were walking on the moon (see Figure 5). Since that time, its share of the federal budget fell from around 4.5 percent to less than 1 percent. As a result, agency leaders believed that they needed to maintain strong relationships with key members of Congress whose states or districts received large portions of the civilian space budget. This in turn required that the agency sustain a large and expensive national infrastructure with more than a dozen major facilities. Although the NASA budget experienced a brief resurgence as EOS was being formulated, this trend reversed itself after only three years. The result produced increasing difficulties for the relevant program managers.[11]

NASA's image as an aged and uninspired bureaucracy clearly troubled

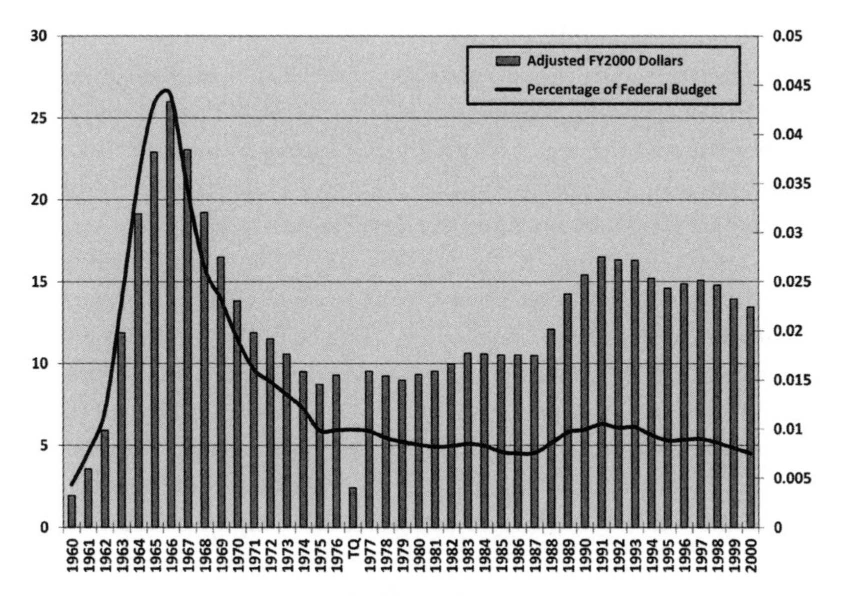


*Figure 5: NASA's Budget, 1960–2000 (billions)*

the Bush White House, which ultimately overrode the space agency's emotive appeals for EOS. Vice President Dan Quayle wrote in his memoir that he thought NASA had projects that were "too unimaginative, too expensive, too big, and too slow." This assessment clearly encompassed agency plans for EOS. Quayle had brought Mark Albrecht to the National Space Council primarily to shake up NASA. As discussed earlier, Albrecht had strong opinions about the space program and particularly of EOS. He believed its high cost profile was the "absolute poster child of the civil space program run amok."[12]

Len Fisk believed that the increased influence of the National Space Council was the predominant reason for changes within the national space program during the early 1990s. This was seen across the agency, but most notably with regard to the SEI. Perhaps the most important lesson learned from the failure of the SEI was that NASA needed competition for ideas from other space policy community actors. In the absence of competition, NASA had not proven itself capable of presenting White House policymakers with a robust suite of policy alternatives. As Vice President Quayle wrote, "The Space Council was now the competition, at least when it came to making policy. . . . The Augustine report told NASA that it had too many people in charge with too little accountability, that it needed to think beyond the shuttle, and that it should not expect the government to build any more orbiters for it." With regard to the SEI, this strategy proved inadequate because the National Space Council failed to act as a truly competitive counterforce—particularly in the absence of more firm control over the development of the initiative. On the other hand, the council was much more effective at providing competition with respect to EOS, which was one of its explicit goals. As Quayle wrote, "While we were willing to back the huge Earth Observing Satellite . . . we wanted the next generation of satellites to be less expensive and faster, to be put into space as a series of components, to reduce the risk of losing everything at once in a faulty launch." The influence of the National Space Council was clearly demonstrated when Science Advisor Allan Bromley requested an NRC review of the USGCRP and EOS in early 1990. Later that year, it was confirmed a second time when the OMB directed the space agency to conduct a more thorough external technical review. As Len Fisk noted later, because the National Space Council derived its authority from Vice President Quayle, it was able to force the space agency "to respond to [its] call for changes in EOS." This fact reinforces the idea that by the early 1990s, NASA was an aging agency in danger of completely losing control of its future destiny within the larger American space program. This situation quite obviously called into question the long-term viability of MTPE/EOS.[13]

*NASA as a Client Bureaucracy*

In his seminal book *Bureaucracy: What Government Agencies Do and Why They Do It*, James Wilson argues that all successful agencies and executives have one thing in common: "They found or maintained the support of key external constituencies. . . . [Therefore, the] principal source of power is a constituency." In creating MTPE/EOS, NASA relied heavily on just such a constituency—the national scientific community. Although the space agency received some assistance from other policy actors interested in enhancing national prestige and promoting international cooperation, the scientific community became the dominant interest group. The conventional wisdom holds that while this type of arrangement has some benefits, it can also result in significant challenges. Wilson writes that when "most or all of the benefits of a program go to some single, reasonably small interest," an agency is forced to engage in client politics. This was certainly the case with MTPE/EOS.[14]

In this book, we have discussed NASA's decades-long struggle for independence from NOAA, the USGS, and other agencies. Wilson considers this type of turf war to be the *external element* of the search for bureaucratic autonomy, where agencies define their jurisdiction. Once the jurisdiction is established, however, there is also an *internal element* of the search for bureaucratic autonomy. With its domain established, an agency can set about determining the means by which it goes about accomplishing its goals. This internal autonomy is a primary characteristic of what Wilson calls a *craft* agency. This type of organization is one whose activities are difficult to observe or understand but whose outcomes are relatively easy to evaluate. The government oversight bodies that approve NASA's budget and watch over its programs cannot easily understand its day-to-day work because it is operating at the technological cutting edge. But the technical and scientific constituency that directly benefits from the space agency's programs can understand these efforts. During the long period when NASA was attempting to win adoption of MTPE/EOS, its plans were tested by a variety of scientific organizations and study groups. In so doing, these external constituencies, which the space agency was dependent upon, were encroaching on the organization's internal autonomy. They were able to do so because they had the technical and scientific expertise necessary to provide constructive criticism and propose different paths forward. This outside influence served as a continuous reminder that because the space agency sometimes required assistance, its autonomy was limited in a very real way. The key question, however, was whether this dynamic worked for or against NASA in its efforts to gain adoption of MTPE/EOS.

Wilson notes that there are important disadvantages to having a strong constituency anchored to an agency's programs. The scientific community encroaching on NASA's internal autonomy could be seen a classic example. As Wilson writes, "Executives who want to influence policy but who define policy largely in terms of what outside constituencies want (or will not denounce) are in an awkward position—more awkward then they sometimes realize." This was never more clear for the space agency than when the scientific community supported significantly redesigning EOS after MTPE had finally been approved. "NASA people were surprised by the scientific objections . . . [they] thought, 'We got you your first flagship mission that will keep you busy for decades, so why are you complaining?'" The answer was surely that important voices within the scientific community were in a position to complain and to demand changes. The reality was that these alternations were not only scientifically sound but also more realistic within the political stream.[15]

Interestingly, the science community's concerns synced nicely with the objectives of the Bush administration. Vice President Quayle not only wanted the National Space Council to compete with NASA on policy matters but also wanted the space agency to adopt the administration's ideas regarding how to run the organization. These ideas were largely embodied by the *cheaper, smaller, and faster* approach that was ultimately championed by Dan Goldin. The release of the Augustine Report provided the perfect opportunity for the space council to trespass upon NASA's autonomy by calling for an external review of the second series of EOS spacecraft, thus guaranteeing that key members of the scientific community would bring a more pragmatic approach to mission planning. The administration hoped this assessment would lead to a stronger focus on research needs than on technical desires. This created a great deal of tension between NASA and the EOS Engineering Review Committee, which had been tasked with conducting the external review. As Edward Frieman recalled later,

> There was quite a struggle ongoing between ourselves and NASA at that point. It wasn't smooth sailing and all hugs and kisses, because clearly NASA and Len Fisk and his group thought the course they [had] embarked on was the correct one. They were talking about huge rockets, huge numbers of instruments on each [satellite]. . . . I guess what I'm really telling you is it was a battle all the way through.[16]

NASA's increasingly complex relationship with the science community, and the client politics that resulted, developed in concurrence with new federal budget priorities. Many commentators have long understood that the single overriding issue in policymaking is the budgetary impact of proposed solutions. As John Kingdon writes, "The budget constitutes a particular kind of problem. . . . Programs, agencies, and professional careers wax and wane according to their budget share. A budget pinch very directly affects both bureaucrats and legislators since the programs in which they have a personal career stake are affected." In 1991, Congress reacted to emerging fiscal constraints by cutting the EOS budget. This was the most effective way for the legislature to exert influence on the program: forcing changes but not eliminating MTPE. As a result, NASA was faced with some very difficult choices. In this situation, the space agency benefited greatly from its strong relationship with the science community within this issue area. In fact, this was the ultimate reason that the organization was able to maintain long-term support for the MTPE/EOS initiative. It was this strong external constituency, and its willingness to speak truth to power, that allowed the successful adoption and perpetuation of the program. The success of the program in turn confounded every expectation with regard to an aging bureaucracy's ability to win support for a new mission. In this situation, an agency's need to play client politics forced it to comport with contemporary political realities. The result was not a burden but a terrific windfall.[17]

## NASA's Path to MTPE/EOS

As we noted in the introduction, this book is the first to describe and analyze the evolution of the MPTE/EOS initiative from its formative years in the 1980s to its political and technical struggles in the 1990s primarily using a policy lens to develop the narrative. As we suggested, this detailed policy history sought to answer several fundamental questions regarding NASA's adoption of the program. Thus, in this section, we return to those questions in an attempt to summarize our most important findings. The first question we posed was as follows: In the early 1980s, why did NASA choose to promote an aggressive earth science research program for inclusion on the national agenda? As we discussed, during the 1970s, the Nimbus and Landsat spacecraft had begun providing data for the initial global studies of the earth. In 1982, Hans Mark asked Richard Goody to assess the space agency's earth science research programs. At the outset, it seemed that Mark wanted no more than an organizing theme for these efforts. He needed a way to package and explain them to protect them from the budget cutters in Reagan's OMB. The study was not intended to justify new spending. However, the scientific community had more on its mind. As the primary client of NASA in this issue area, it worked through the Goody committee to propose a bold new interagency research program. This initiative would be led by NASA and would study the impact of human activity on global processes such as climate. The scientific community continued to push this agenda item in various reports produced by the NRC. In particular, both the Space Studies Board and the Friedman committee called for a more thorough and expansive study of the earth.

Despite these efforts coordinated by key members of the scientific community, NASA stayed true to Mark's original intent and simply used the idea of global change as an organizing theme for its existing research. Thus, its global habitability initiative was not really an effort on the space agency's part to justify additional spending on a more thorough and expansive study of the earth but was just an effort to protect the funding it already had. This became quite clear in its presentations at the Unispace conference in August 1982, which were seen not as a genuine effort to kick-start an important new program but rather as an attempt to grab power within the contemporary earth science research arena. The reaction was swift and brutal. If NASA wanted to take on a leadership role, global habitability was the wrong way to gain the support of the scientific community.

This leads us to the second question: Why did NASA, which had not been keen on battling for a new initiative in the first place, spend the remainder of the decade pursuing it? The reason was that the program became linked to the most important human spaceflight initiative of the day. In 1982, Jim Beggs had initiated a two-year process to gain President Reagan's approval to build a space station. To build up the case for this program, Beggs wanted ideas for how a station could support science. Burt Edelson suggested that a human-tended station could be complemented by a flotilla of scientific satellites—some co-orbiting with it and some in polar orbit. These satellites would be launched on the space shuttle and serviced by astronauts, creating a *space station complex* that provided a large array of scientific capabilities. The idea of large, complex structures in space appealed to Edelson, who was an engineer by training and believed the agency should tackle big engineering projects to develop advanced technology. He had one specific example in mind that could be carried out as part of a space station complex. He called it System-Z.

System-Z was a series of large polar-orbiting satellites that would enable earth science research. They would be serviced by astronauts and would operate for more than a decade. These spacecraft would enable global change research, as advocated by the Goody committee and the NRC committees in recent years. Such an initiative would accomplish two things. First, it would give the earth science community its first flagship mission. Second, it would give the human spaceflight program an ally in the earth science research community and bolster support for the space station. In other words, Edelson believed it would create a win-win synergy between two previously separate programs. The result was that rather than merely having an organizing theme for earth science research, NASA had a plan for a more advanced and thorough research program. Both Beggs and Mark endorsed this idea and System-Z, later renamed the Earth Observing System (EOS), became a central part of NASA's space station plans. This entire approach was somewhat thrown into disarray, however, with the shuttle *Challenger* accident. MTPE/EOS was ultimately decoupled from the station program, and NASA sought to gain approval for the program on its own merits during the last few years of the decade.

Thus we arrive at our third question: In 1989, how did the initiative reach the national agenda and gain formal adoption? During the years leading up to the decision, several overlapping events converged favorably. Much of this convergence involved organizing the scientific community to provide support for the program, which became crucial, as we have seen. In the three years prior to the final approval of the undertaking, the Bretherton committee, the Eddy committee, and the Space Studies Board all produced reports that outlined research goals. As these studies were being carried out, leading scientists were increasingly testifying before Congress with regard to the dangers posed by air pollution, ozone depletion, and global warming. In some respects, Sally Ride's placement of MTPE on an equal footing with human spaceflight and space science initiatives solidified it as an agency priority. When George H. W. Bush was elected to the presidency, he wanted bold new initiatives in the area of environmental research. Both the USGCRP and MTPE fit the bill. From a White House perspective, these programs had been well thought out and had the support of the scientific community. Bush was also confident that the Democratic Congress would back both efforts. Thus, when the policy window opened, Richard Truly was quickly able to gain approval for MTPE/EOS.

The story did not end there, however, which allows us to consider the final question: During the bureaucratic battles that ensued in the 1990s, how did NASA maintain support for the program as political and budgetary priorities changed? From the outset, NASA had considered large satellites the proper technical approach for EOS. Agency leaders liked the idea of a satellite constellation that would be serviced by the space shuttle, which allowed for upgrading instruments as new remote sensing technologies became available. This approach meant that EOS would be almost as complex as a human-tended space station. NASA never wavered from this big satellite approach during the 1980s, even as important voices within the scientific community began expressing concerns. The agency saw small satellites only as complementary tools. Shortly after receiving approval for a new start, however, the Bush administration and key appropriators came to realize that there was a need to scale back the technical scope of the mission. Simply put, EOS was too expensive. Furthermore, the original concept was thought to be unresponsive to the needs of the scientific community. Changing EOS proved to be painful. As an administration official said later, "Unfortunately, we were already pregnant with the original behemoth platforms, which interestingly were tied to the early designs of the Space Station, which proved too cumbersome and difficult to easily change in direction. Hence, we had a decade of much turmoil."[18] As budgets were steadily cut, the space agency was forced to be smarter in how it approached the program, beginning with Edward Frieman's restructuring work and continuing with the internal rescoping and the external reshaping exercises after Dan Goldin arrived with his *faster, better, cheaper* philosophy. Although few within NASA at the time recognized it, this process ensured MTPE/EOS's survival at a time when an environmental monitoring program had limited support in the White House or Congress. Thus, the space agency was able to retain this critical Big Science program, which ultimately became a key contributor of needed data within the increasingly important debate about earth system science and global change.

## EOS Contributions to Earth System Science and Global Change Research

During the past two decades, NASA has continued to battle to maintain funding for EOS. Particularly during the George W. Bush and Trump administrations, the relevance of a scientific endeavor closely linked with climate change came under increased scrutiny. For the most part, however, the space agency has been able to maintain a surprisingly robust earth science program. NASA has launched twenty-six different EOS satellites, and more are planned for the future. As Ellen Gray and Patrick Lynch write, during its operational life EOS has "produced a more comprehensive look at Earth from space than any other period in history. At a time when our planet is undergoing critically important changes, this global view offers not only stunning imagery but also vitally important information about how Earth is changing." The amount of data that EOS satellites have gathered in less than two decades is mind-boggling, and earth and global change scientists are using it every day to increase our collective knowledge of how our planet operates. The number of new discoveries is far too large to discuss comprehensively, but it seems worthwhile to highlight a few important scientific accomplishments of EOS.[19]

NASA has been at the forefront of research on the rapid changes impacting the earth's polar regions. One aspect of these studies has been following the decay of the antarctic ice shelves, which could eventually threaten rapid sea-level increases. In 2002, the *Terra* satellite monitored the collapse of the Larsen B Ice Shelf, which was larger than New Hampshire and Vermont combined, over the period of several months. This provided scientists with data regarding the processes and pace of the deterioration of the larger West Antarctica Ice Sheet—the disintegration of this ice sheet would lead sea levels to rise at least 10 feet worldwide. The result would be the need to protect or evacuate millions of acres of landmass, creating a refugee crisis with the potential to destabilize the entire global community.[20]

At the opposite end of the planet, NASA has also been heavily engaged in examining declines in the amount of summer sea ice in the Arctic. Maintaining this sea ice is critical because it reflects huge amounts of solar radiation back into space. When the polar ice cap shrinks, sun energy reaches the ocean below and heats it significantly. This in turn creates a worrisome feedback loop that could lead to greater sea-ice losses and thus more atmospheric heating. Temperatures are already rising at the poles two to three times faster than on the rest of the planet, so the continued contraction of the polar ice cap is a matter of great concern. Both the *Terra* and *Aqua* satellites "have now been measuring the amount of solar radiation absorbed by the Earth since the year 2000. And in the Arctic this absorption has increased by 5 percent in that time—a number that may not seem like much, except when you consider that the absorption rate has been essentially flat across the rest of the globe, and no other region on Earth shows a pattern of change." It is also worth noting that this is further disturbing because in addition to the polar ice cap, the Greenland Ice Sheet is almost entirely located within the Arctic Circle. As is the case at the southern pole, the loss of this northern ice sheet could ultimately lead sea levels to rise twenty-five feet worldwide. Recent studies have suggested that the Greenland Ice Sheet is melting far faster than previously believed, which is driving continued sea-level rises. The EOS-generated data set has provided interested policymakers with compelling evidence that the risks of climate change–related sea-level increases are very serious.[21]

It has long been known that carbon dioxide emissions fluctuate every year, which can be seen in charts showing increased concentrations in the global atmosphere over time. However, it was never possible to map this phenomenon across the entire planet. The AIRS instrument on the *Aqua* satellite has filled this gap by using its data to create a visualization of how greenhouse gases are dispersed around the globe and how they oscillate with the seasons. Similarly, NASA now has access to data examining the global vegetation cycle. EOS satellites have provided detailed photographs of where plants and trees grow, which is critical to understanding "how much carbon dioxide plants are absorbing from the atmosphere during photosynthesis." Yet another type of important data that EOS collects examines the burning of global forests, due to both wildfires and human efforts to make way for additional farmland. Data provided by *Terra* and *Aqua* revealed that "each year about one third of Earth's land surface is touched by fire. [This has] revolutionized scientists' understanding of where fires occur and how they affect ecosystems, carbon released into the atmosphere that contributes to climate change, and air quality that affects human health." Each of these efforts has led to transformations in the way that earth scientists understand the global carbon cycle that would not have been possible without access to satellite observations.[22]

One of the last frontiers of climate science is to better understand the role that aerosols and clouds play in global change because both aerosols and clouds can either reflect or trap solar radiation in different circumstances. When they reflect sun energy, they reduce global warming. When they trap it, they promote global warming. *Terra* and *Aqua*, working with additional satellites flying in tight formations, have been compiling a continuous long-term record of clouds and aerosols. The hope is that over time, this data will help climate scientists understand their overall impact on global change.[23]

In late 2015, the Space Studies Board began the process of assembling its second decadal survey of space-based earth observations. The first iteration was completed in 2007 and was a remarkable achievement because earth scientists from a wide variety of disciplinary backgrounds reached consensus on objectives within this arena. The follow-up effort began by assembling a steering committee responsible for drafting a final report and five working groups tasked with providing recommendations within specified study areas. After more than a year of research, the Space Studies Board published its seven-hundred-page report in January 2018. The primary focus of this analysis was to provide program managers and policymakers with strategic guidance for the forthcoming decade, but the study team also included an assessment of the accomplishments of the space-based earth observation program.[24]

The report opened by arguing that the ability to systematically monitor the planet from space has "enabled societal applications that provide tremendous value to individuals, businesses, the nation, and the world. Such applications are growing in breadth and depth, becoming an essential information infrastructure element for society as they are integrated into people's daily lives." NASA, NOAA, the USGS, and other federal agencies have made this possible because of a persistent commitment to developing and operating satellite and information systems. The study team contended that progress made during the past decade was considerable, pointing to three areas where different actors were making use of space-based observations:

- *Individuals.* Greater access to this information has helped us as individuals by placing at our fingertips a wide variety of vital information about the world around us, helping each of us make important decisions. Examples range from minute-by-minute weather information to satellite images that allow us all to explore and navigate in our hometowns. . . .
- *Businesses.* Scientific discoveries and the resulting applications have helped us advance business interests, such as making our agriculture more productive, our energy use more efficient, and our transportation more reliable. . . .
- *Society.* Being able to observe the Earth in new ways has helped us prosper as a society. Our revolutionary ability to view the world as a whole from space allows us to watch the natural course of rivers and forests change, to observe changes in our climate, to discern our role within those and other changes, to understand the risks and benefits of our actions and inactions with regard to our planet, and to apply the resulting knowledge. This expanded perspective has positioned us to benefit from the economic opportunities it creates, increased our resilience to the environment's risks, and inspired citizens and nations everywhere with the wonder of Earth's scientific challenges.

EOS data and the scientific community have made it possible to understand to a greater extent our planet as one large interconnected natural and social order.[25]

The Space Studies Board found that during the past decade, EOS has contributed to substantial advances in knowledge concerning earth system science. Because so much had happened in ten years, the study team decided to highlight just fifteen areas in which considerable progress had been made:

- Advancing weather prediction skills
- Understanding of air and sea fluxes of sensible and latent heat
- Tracking extreme precipitation to reduce disaster risk
- Enhancing monitoring to support improvements in U.S. air quality
- Tracking sea-level rise and its sources
- Monitoring and understanding stratospheric ozone
- Increasing global availability of satellite-based emergency mapping
- Revolutionizing our understanding of marine ecosystems
- Quantifying worldwide emissions and concentrations of air pollutants
- Using satellite data in health impact assessment
- Monitoring land-use change due to both human and natural causes
- Tracking variations in ocean plankton and land vegetation
- Seeing the rain formation process for the first time on a global scale
- Examining cloud feedbacks contributing to the decadal cooling in the eastern tropical Pacific
- Observing a slowdown in sea-level rise associated with flooding in Australia

One final finding of the Space Studies Board in its decadal survey is that beyond scientific developments, EOS and other space-based observation systems have also had important and unplanned spinoff impacts within the commercial sector. The committee wrote, "This productive coupling between curiosity-driven science and applications-driven research is a hallmark of Earth system science." While this is certainly a happy result, national leaders should remain primarily focused on the scientific contributions of these programs.[26]

In 2017, the National Academies of Sciences, Engineering, and Medicine completed an assessment of the contributions that the overarching USGRCP has made to human knowledge about the forces that are transforming the global climate. The committee tasked with carrying out this examination listed four principal areas in which the agencies involved in this effort have made extraordinary progress in "building scientific knowledge and making it useful." The first success on the committee's list was global observation systems. While much of the data being used by climate researchers is gathered in situ, the report makes clear the outsized role played by EOS satellites and other orbiting platforms. As the study team wrote, "Earth system science cannot be done on a bench in a laboratory—it is an observational science where the laboratory is the planet. Without continuous global observations that are quality controlled for long time series analysis, it is impossible to make progress on documenting and understanding global environmental changes." The committee highlighted NASA's decision to champion the creation of a space-based observational system capable of providing earth scientists with the systematic data necessary to advance knowledge about some of the most important questions facing society in the modern era. The result has been the ability to measure and/or calculate "wind speeds and circulation patterns, atmospheric concentrations of $CO_2$ and other important chemical constituents, land-cover change, ice sheets, cloud properties, ocean salinity, sea ice, and the biosphere's net primary productivity on both land and ocean."[27]

NASA's contribution in providing global observational data also played an important role in two of the other major accomplishments highlighted by the National Academies. The committee commended the USGCRP for promoting significant advances in earth system modeling, suggesting that this is an "indispensable tool for better understanding Earth system science and in generating information needed by decision makers at all levels." While much of the credit in this arena must obviously go to the computer engineers and earth scientists who develop new systems for building these models, most of the resulting outputs would not be possible without the initial input of data from EOS satellites. As discussed above, another area of rapid progress has been a more sophisticated grasp of carbon-cycle science, which has been critical because of the role that carbon plays as a control mechanism for the global climate. Once again, EOS data has been important for research aimed at providing an improved comprehension of the "rates and causes of both carbon fluxes to the atmosphere . . . and carbon sequestration in land and ocean ecosystems . . . [which] is essential for developing policies to manage climate change." While it is clear from reading the report that much more needs to be done in the future, USGRCP and EOS have already made noteworthy contributions to advancing human knowledge about earth system science and have provided an important tool for policymakers interested in developing solutions to emerging environmental problems.[28]

Since its inception, EOS has fundamentally changed the way we view our biosphere. It has proved critical in early efforts to take more aggressive action to address both regional and global environmental challenges. It has become increasingly evident that tremendous benefits have accrued from the decisions made by NASA leaders at the end of the past century, with the support and at the urging of the scientific community, to push forward a program to provide space-based observations of our home world. Combined with in situ observations, space-based satellites and instruments have proven to be crucially important. Environmentalists, earth scientists, and national leaders must maintain a commitment to EOS in the years and decades to come. The very ability of humanity to thrive on this globe is at stake. Let us move forward with confidence that government-supported science can provide us with the tools necessary to better understand our planet and a belief that we can use this knowledge to build a better future for all.

# Notes

## Introduction

1. Howard McCurdy, *Space and the American Imagination* (Baltimore, MD: Johns Hopkins University Press, 2011), 11–32.

2. Ray A. Williamson and Roger D. Launius, "Rocketry and the Origins of Space Flight," in *To Reach the High Frontier: A History of U.S. Launch Vehicles*, eds. Roger D. Launius and Dennis R. Jenkins (Lexington: University of Kentucky Press, 2002), 34.

3. Williamson and Launius, "Rocketry and the Origins of Space Flight," 34.

4. William Burroughs, *This New Ocean: The Story of the First Space Age* (New York: Modern Library, 1999), 36–73; Milton Lehman, *This High Man* (New York: Farrar, 1963).

5. Williamson and Launius, "Rocketry and the Origins of Space Flight," 42–43; Burroughs, *This New Ocean*, 94–107.

6. Asif A. Siddiqi, "Korolev, Sputnik, and The International Geophysical Year," NASA History Office, accessed 21 November 2002, tinyurl.com/yczzyoj5.

7. Siddiqi, "Korolev, Sputnik, and The International Geophysical Year."

8. Williamson and Launius, "Rocketry and the Origins of Space Flight," 44–49.

9. Williamson and Launius, 44–49.

10. Williamson and Launius, 49–52.

11. Williamson and Launius, 49–52; Lomansk C. Green, *Vanguard: A History* (Washington, DC: Government Printing Office, 1970), 17–48.

12. Martha W. George, "The Impact of Sputnik I: Case-Study of American Public Opinion at the Break of the Space Age, October 4, 1957," NASA Historical Note 22, 15 July 1963; Walter McDougall, . . . *the Heavens and the Earth: A Political History of the Space Age* (New York: Basic Books, 1985), 112–141; George Price, "Arguing the Case for Being Panicky" *Time* (16 November 1957), 125–128; "Senators Lash Defense Policy," *Washington Post* (7 October 1957); Robert A. Divine, *The Sputnik Challenge: Eisenhower's Response to the Soviet Satellite* (Oxford, UK: Oxford University Press, 1993), xv–7.

13. McDougall, . . . *the Heavens and the Earth*, 3–4.

14. "Statement by the President Summarizing Facts in the Development of an Earth Satellite by the United States, 9 October 1957," in *Public Papers of the*

*Presidents of the United States: Dwight David Eisenhower* (Washington, DC: Government Printing Office, 1957), 733–735; "The President's News Conference, 9 October 1957," in *Public Papers of the Presidents of the United States: Dwight David Eisenhower* (Washington, DC: Government Printing Office, 1957), 719–732; Gallup Poll, 8 January 1958.

15. Divine, *The Sputnik Challenge*, 153–156.

16. "Special Message to the Congress Relative to Space Science and Exploration, 2 April 1958," in *Public Papers of the Presidents of the United States: Dwight David Eisenhower* (Washington, DC: Government Printing Office, 1958), 269–273.

17. McDougall, . . . *the Heavens and the Earth*, 177–194.

18. McDougall, . . . *the Heavens and the Earth*, 167–221; Lloyd Swenson, James Grimwood, and Charles Alexander, *This New Ocean: A History of Project Mercury* (Washington, DC: Government Printing Office, 1998); Linda Neumann Exell, *NASA Historical Data Book,* Volume 2: *Programs and Projects 1958–1968* (Washington, DC: NASA SP-4012, 1988), 124; National Aeronautics and Space Administration, *Objectives and Basic Plan for the Manned Satellite Project* (Washington, DC: Government Printing Office, 1958).

19. McDougall, 218–220 and 301–324; Walt Rostow to John Kennedy, "A Democratic Strategy for 1960," 2 January 1960, John F. Kennedy Presidential Library, Boston, MA.

20. John M. Logsdon, *Decision to Go to the Moon: Project Apollo and the National Interest* (Chicago: University of Chicago Press, 1970); Ad Hoc Committee on Space, "Report to the President-Elect, 10 January 1961," in *Exploring the Unknown—Selected Documents in the History of the U.S. Civil Space Program*, Volume 1: *Organizing for Exploration*, ed. John Logsdon, Linda J. Lear, Jannelle Warren-Findley, Ray A. Williamson, and Dwayne A. Day (Washington, DC: NASA History Office, 1995), 416–423; Air Force Space Study Committee, "Gardner Report," 20 March 1961, NASA History Office, Washington, DC; McDougall, 309–316; Burroughs, *This New Ocean*, 311–317; Michael Beschloss, "Kennedy and the Decision to Go to the Moon," in *Spaceflight and the Myth of Presidential Leadership*, ed. Roger D. Launius and Howard E. McCurdy (Chicago: University of Illinois Press, 1997), 55–57.

21. John Kennedy to Lyndon Johnson, 20 April 1961, in *Exploring the Unknown*, Volume 1, 424; Logsdon, *Decision to Go to the Moon*, 123–125; James Webb and William McNamara to Lyndon Johnson, "Recommendations for Our National Space Program: Changes, Policies, Goals," 8 May 1961, in *Exploring the Unknown*, Volume 1, 439–452; John Kennedy, "Urgent National Needs," 25 May 1961, in *Exploring the Unknown*, Volume 1, 453–454; Courtney Brooks, James Grimwood, and Loyd Swenson, *Chariots for Apollo: A History of Manned Lunar Spacecraft* (Washington, DC: Government Printing Office, 1979), epilogue.

22. Joseph Pelton, "The History of Satellite Communications," in *Exploring*

*the Unknown—Selected Documents in the History of the U.S. Civil Space Program*, Volume 3: *Using Space*, ed. John Logsdon, Roger D. Launius, David H. Onkst, and Stephen J. Garber (Washington, DC: NASA History Office, 1998), 1–11.

23. John Nagle and John Logsdon, "Space Science: Origins, Evolution, and Organization," in *Exploring the Unknown—Selected Documents in the History of the U.S. Civil Space Program*, Volume 5: *Exploring the Cosmos*, ed. John Logsdon, Amy Paige Snyder, Roger D. Launius, Stephen J. Garber, and Regan Anne Newport (Washington, DC: NASA History Office, 2001), 1–15.

24. Amy Paige Snyder, "NASA and Planetary Exploration," in *Exploring the Unknown*, Volume 5, 271–298.

25. Snyder, "NASA and Planetary Exploration," 271–298.

26. Snyder, 271–298.

27. Snyder, 271–298.

28. Leon Jaroff, "Spinning Out of Orbit," *Time* (August 6, 1990), 26–27; John Logsdon, "The Evolution of U.S. Space Policy and Plans," in *Exploring the Unknown*, Volume 1, 382–388; T. A. Heppenheimer, *The Space Shuttle Decision: NASA's Search for a Reusable Launch Vehicle* (Washington, DC: NASA History Office, 1999), ix–x.

29. Jaroff, "Spinning Out of Orbit"; Logsdon "The Evolution of U.S. Space Policy and Plans".

30. Logsdon, "The Evolution of U.S. Space Policy and Plans," 390–392.

31. Michael Cohen, James March, and Johan Olsen, "A Garbage Can Model of Organizational Choice," *Administrative Science Quarterly* (March 1972): 1–25; Wayne Parsons, *Public Policy: An Introduction to the Theory and Practice of Policy Analysis* (Cheltenham, UK: Edward Elgar Publishing, 1996), 192–193.

32. John W. Kingdon, *Agendas, Alternatives, and Public Policies* (New York: HarperCollins College Publishers, 1995), 86–89.

33. Kingdon, 26–42 and 231–240; Thor Hogan, *Mars Wars: The Rise and Fall of the Space Exploration Initiative* (Washington: NASA History Series, 2007), 2–4 and 138–142.

34. Hogan, *Mars Wars*, 2–4 and 138–142.

35. Hogan, 2–4 and 138–142.

36. Hogan, 2–4 and 138–142.

37. Hogan, 2–4 and 138–142.

38. Frank R. Baumgartner and Bryan D. Jones, *Agendas and Instability in American Politics* (Chicago: University of Chicago Press, 1993), 3–24.

39. Anthony Downs, *Inside Bureaucracy* (Boston, MA: Little, Brown and Company, 1966), quotation on 1.

40. Downs, *Inside Bureaucracy*, chapter 2.

41. James Q. Wilson, *Bureaucracy: What Government Agencies Do and Why They Do It* (New York: Basic Books, 1989).

42. Erik Conway, *Atmospheric Science at NASA: A History* (Baltimore, MD: Johns Hopkins University Press, 2008).

43. Richard LeRoy Chapman, *A Case Study of the U.S. Weather Satellite Program: The Interaction of Science and Politics* (PhD diss., Syracuse University, 1967).

44. Janice Hill, *Weather from Above: America's Meteorological Satellites* (Washington, DC: Smithsonian Institution Press, 1991).

45. Pamela E. Mack, *Viewing the Earth: The Social Construction of the Landsat Satellite Program* (Cambridge, MA: MIT Press, 1990).

46. John H. McElroy and Ray A. Williamson, "The Evolution of Earth Science Research from Space: NASA's Earth Observing System," in *Exploring the Unknown—Selected Documents in the History of the U.S. Civil Space Program*, Volume 6: *Space and Earth Sciences*, ed. John M. Logsdon (Washington, DC: NASA History Series, 2004), 441–473.

47. Edward S. Goldstein, "NASA's Earth Science Program: A Case Study" (PhD diss., George Washington University, 2004).

48. National Research Council, Committee on Earth Studies, *Earth Observation from Space: History, Promise, and Reality* (Washington, DC: National Academy Press, 1996).

49. Richard Goody, *Global Change: Impacts on Habitability* (Pasadena, CA: Jet Propulsion Laboratory, 1982).

50. National Research Council, Space Studies Board, *A Strategy for Earth Science from Space* (Washington, DC: National Academy Press, 1982).

51. National Research Council, Steering Committee on an International Geosphere-Biosphere Program, *Toward an International Geosphere-Biosphere Program: A Study of Global Change* (Washington, DC: National Academy Press, 1983).

52. Sally Ride, *Leadership and America's Future in Space* (Washington, DC: NASA, 1987).

53. NASA Advisory Council, Earth System Science Committee, *Earth System Science: A Closer Look: A Program for Global Change* (Washington, DC: NASA, 1988).

54. Committee on Earth Sciences, *Our Changing Planet: A U.S. Strategy for Global Change Research, a Report by the Committee on Earth Sciences to Accompany the President's Fiscal Year 1990 Budget* (Washington, DC: Office of Science and Technology Policy, 1989)

55. *Report of the Advisory Committee on the Future of the U.S. Space Program* (Washington, DC: NASA, 1990).

56. *Report of the Earth Observing System (EOS) Engineering Review Committee* (Washington, DC: NASA, 1991).

57. Howard McCurdy, *The Space Station Decision: Incremental Politics and Technological Choice* (Baltimore, MD: Johns Hopkins University Press, 1990).

58. Henry W. Lambright, "The Ups and Downs of Mission to Planet Earth," *Public Administration Review* 54, no. 2 (March/April 1994): 97.

59. Edward Edelson, "Laying the Foundation," *Mosaic* 19, no. 314 (Fall/Winter 1988), 6–7.

60. David Kennedy, "The U.S. Government and Global Environmental Change Research: Ideas and Agendas," Case C16-92-1121.0, John F. Kennedy School of Government, Harvard University (1992); David Kennedy, "This Far and No Further: The Rise and Fall of the Committee on Earth and Environmental Sciences," Case C16-92-1122.0, John F. Kennedy School of Government, Harvard University (1992).

61. Roger A. Pielke Jr., "Completing the Circle: Global Change Science and Usable Policy Information" (PhD diss., University of Colorado, 1994).

## Chapter 1. A Global View of the Earth

1. Pamela Mack, *Viewing the Earth: The Social Construction of the LANDSAT Satellite System* (Cambridge, MA: MIT Press, 1990), 32–33; James B. Campbell and Randolph Wynne, *Introduction to Remote Sensing* (New York: Guilford Press, 2011).

2. Campbell and Wynne, *Introduction to Remote Sensing*, 11; Douglas Aircraft Corporation, "Preliminary Design of an Experimental World-Circling Spaceship," Report No. SM11827, 2 May 1946, available in *Exploring the Unknown—Selected Documents in the History of the U.S. Civil Space Program*, Volume 3: *Using Space*, ed. John Logsdon et al. (Washington, DC: Government Printing Office, 1998); Walter McDougall, . . . *the Heavens and the Earth: A Political History of the Space Age* (Baltimore, MD: Johns Hopkins University Press, 1985), 102–105.

3. McDougall, . . . *the Heavens and the Earth*, 108–110, quotation on 108; Paul Kecskemeti, "The Satellite Rocket Vehicle: Political and Psychological Problems," RAND RM-567, 4 October 1950.

4. McDougall, . . . *the Heavens and the Earth*, 111; Samuel Willard Crompton, *Sputnik/Explorer I: The Race to Conquer Space* (New York: Chelsea House Publishers, 2007), 23–34.

5. Pamela E. Mack and Ray A. Williamson, "Observing the Earth from Space," in *Exploring the Unknown*, Volume 3, 155–156.

6. Mack and Williamson, "Observing the Earth from Space," 155–156; Linda Neumann Exell, *NASA Historical Data Book*, Volume 2: *Programs and Projects, 1958–1968* (Washington, DC: NASA SP-4012, 1988), Table 3-186, "Tiros 1 Characteristics," 352. The Weather Bureau was founded in 1891 in the Department of Agriculture by the transfer of the civilian meteorological service from the U.S. Army Signal Corps. The Weather Bureau first became responsible for providing weather information to other parts of the federal government in 1926, when the

Air Commerce Act made it the weather service provider for civil aviation. In 1940, the Weather Bureau was transferred to the Department of Commerce.

7. Janice Hill, *Weather from Above: America's Meteorological Satellites* (Washington, DC: Smithsonian Institutional Press, 1991), 15–21.

8. Mack and Williamson, "Observing the Earth from Space," 155–177; Hill, *Weather from Above*, 21–22.

9. Mack and Williamson, "Observing the Earth from Space," 155–177; Hill, *Weather from Above*, 21–22.

10. Hugh L. Dryden (NASA) and Luther H. Hodges (Weather Bureau), "Basic Agreement between U.S. Department of Commerce and the National Aeronautics and Space Administration Concerning Operational Meteorological Satellite Systems," 30 January 1964, in *Exploring the Unknown*, Volume III, 206.

11. Paul McCeney, "Applications Technology Satellite Program," in *Trends in Communications Satellites*, ed. D. J. Curtin (New York: Pergamon Press, 1979), 299–326.

12. National Research Council Committee on the NASA-NOAA Transition from Research to Operations, *Satellite Observations of the Earth's Environment: Accelerating the Transition of Research to Operations* (Washington, DC: National Academy Press, 2003).

13. Erik Conway, *Atmospheric Science at NASA: A History* (Baltimore, MD: Johns Hopkins University Press, 2008), 39–93.

14. Conway, *Atmospheric Science at NASA*, 39–93; Hill, *Weather from Above*, 26.

15. Conway, *Atmospheric Science at NASA*, 39–93.

16. Mack, *Viewing the Earth*, 32–33, quotation on 38; Donald T. Lauer, Stanley A. Morain, and Vincent V. Salomonson, "The LANDSAT Program: Origins, Evolution, and Impacts," *Photogrammetric Engineering and Remote Sensing* 63 (July 1997): 832.

17. Paul Lowman, Jr., "LANDSAT and Apollo: The Forgotten Legacy," *Photogrammetric Engineering and Remote Sensing* 65 (October 1999): 1143; Mack, *Viewing the Earth*, 40; T. M. Lillesand, Ralph Kiefer, and Jonathan Chipman, *Remote Sensing and Image Interpretation* (Hoboken, NJ: John Wiley, 1994), 298, 380, and 398.

18. Mack, *Viewing the Earth*, 46–55.

19. Lowman, "LANDSAT and Apollo," 1144; Mack and Williamson, "Observing the Earth from Space," 155–177; Office of the Secretary, U.S. Department of the Interior, "Earth's Resources to Be Studied from Space," News Release, 21 September 1966, in *Exploring the Unknown*, Volume 3 , 244.

20. Mack, *Viewing the Earth*, 56–63, quotation on 65.

21. Mack, 60–61.

22. Mack, 61–65, quotation on 86.

23. Congressional Research Service, "The Future of the Land Remote Sensing Satellite System (Landsat)," 13 April 1989, NASA History Office archives.

24. Mitchell Waldrop, "Imaging the Earth (I): The Troubled First Decade of Landsat," *Science*, 26 March 1982, 1600–1603; Pamela Mack, "LANDSAT and the Rise of Earth Resources Monitoring," in *From Engineering Science to Big Science: The NACA and NASA Collier Trophy Research Project Winners*, ed. Pamela Mack (Washington, DC: NASA History Series, 1998), quotation on 235.

25. Lauer, Morain, and Salomonson, "The Landsat Program," 831–838.

26. Ray Williamson, "The LANDSAT Legacy: Remote Sensing Policy and Development of Commercial Remote Sensing," *Photogrammetric Engineering and Remote Sensing* 63 (July 1997): 877.

27. Daniel R. Glover, "NASA Experimental Communication Satellites—1958–1995," in *Beyond the Ionosphere: Fifty Years of Satellite Communication*, ed. Andrew Butrica (Washington, DC: NASA History Series, 2013), chapter 6. NOAA also decided to procure a satellite based on the upgraded ATS design in order to add a geosynchronous satellite to its operational capabilities and complement the TOS series. NOAA called the satellite the Geostationary Operational Environmental Satellite (GOES). GOES is still the name of NOAA's geostationary weather satellite system today.

28. H. F. Eden, B. P. Elero, and J. N. Perkins, "Nimbus Satellites: Setting the Stage for Mission to Planet Earth," *EOS*, 29 June 1993, 285; Conway, *Atmospheric Science at NASA*, 39–93; Hill, *Weather from Above*, 36.

29. Conway, *Atmospheric Science at NASA*, 39–93; I. S. Haas and R. Shapiro, "The Nimbus Satellite System: Remote Sensing R&D Platform of the 1970s," in *Monitoring Earth's Ocean, Land, and Atmosphere from Space—Sensors, Systems, and Applications*, ed. Abraham Schnapf (Reston, VA: American Institute of Aeronautics and Astronautics, 1985), 71–93; Ellis Remsberg, "Remote Measurement of Pollution—A 40-Year Langley Retrospective, Part II: Aerosols and Clouds," NASA Langley Research Center, 1 May 2012.

30. Eden, Elero, and Perkins, quotation on 285; Haas and Shapiro, 71–93.

31. William Sheehan, *The Planet Mars: A History of Observation and Discovery* (Tucson, AZ: University of Arizona Press, 1996), 185–216; John Noble Wilford, *Mars Beckons: The Mysteries, the Challenges, the Expectations of Our Next Great Adventure in Space* (New York: Vintage Books, 1990), 53–61.

32. Carl Sagan and George Mullen, "Earth and Mars: Evolution of Atmospheres and Surface Temperatures," *Science*, 7 July 1972, 52–56.

33. Sagan and Mullen, "Earth and Mars," 52–56; Conway, *Atmospheric Science at NASA*, quotation on 112.

34. James Lovelock, *Gaia: A New Look at Life on Earth* (Oxford, UK: Oxford University Press, 1974); James Lovelock and Lynn Margulis, "Biological Regula-

tion of the Earth's Atmosphere," *Icarus* 21 (1974): 471; Conway, *Atmospheric Science at NASA*, quotation on 116.

35. Erik Conway, *High Speed Dreams: NASA and the Technopolitics of Supersonic Transportation* (Baltimore, MD: Johns Hopkins University Press, 2005), 166–169.

36. Conway, *High Speed Dreams*, 166–169, quotations on 166 and 167.

37. "SST Cleared on Ozone," *Washington Post*, 25 January 1975; Conway, *Atmospheric Science at NASA*, 135.

38. Henry Lambright, *NASA and the Environment: The Case of Ozone Depletion* (Washington, DC: NASA SP-2005-4538, 2005), 7–8, quotations on 7 and 8; Conway, *High Speed Dreams*, 168.

39. Henry Lambright, "NASA, Ozone, and Policy-Relevant Science," *Research Policy* (1995): 747–760, quotation on 750.

## Chapter 2. Evolution of an Idea

1. Mitchell Waldrop, "Reagan Fills Top Posts at NASA," *Science*, 8 May 1981, 646; Howard McCurdy, *The Space Station Decision: Incremental Politics and Technological Choice* (Baltimore, MD: Johns Hopkins University Press, 1990), 36; Executive Office of the President, *America's New Beginning: A Program for Economic Recovery*, 18 February 1981.

2. Hans Mark, *The Space Station: A Personal Journey* (Durham, NC: Duke University Press, 1987), 28–29 and 90–115; Waldrop, "Reagan Fills Top Posts at NASA," 646.

3. Mark, *The Space Station*, 121; McCurdy, *The Space Station Decision*, 39 and 50; Adam L. Gruen, *The Port Unknown: A History of the International Space Station Freedom Program*, unpublished manuscript (1992), 62–65, NASA History Office, Space Station History Collection, Washington, DC.

4. Pamela Mack and Ray Williamson, "Observing the Earth from Space," in *Exploring the Unknown—Selected Documents in the History of the U.S. Civil Space Program*, Volume 3: *Using Space*, ed. John Logsdon et al. (Washington, DC: NASA History Series, 1998), 162.

5. Amy Paige Snyder, "NASA and Planetary Exploration," in *Exploring the Unknown—Selected Documents in the History of the U.S. Civil Space Program*, Volume 5: *Exploring the Cosmos*, ed. John Logsdon et al. (Washington, DC: NASA History Series, 2001), 290.

6. James Beggs to David Stockman, 29 September 1981, in *Exploring the Unknown*, Volume 5; Snyder, 290.

7. Erik Conway, *Atmospheric Science at NASA: A History* (Baltimore, MD: Johns Hopkins University Press, 2008), 91–92.

8. National Research Council, Space Studies Board, *A Strategy for Earth Science from Space, Part 1: Solid Earth and Oceans* (Washington, DC: National Academy Press, 1982).

9. National Research Council, Space Studies Board, *A Strategy for Earth Science from Space*, 21–98, quotations on 21.

10. National Research Council, Space Studies Board, *A Strategy for Earth Science from Space*, quotations on 86, 89, 96, and 98.

11. Richard Goody, "Observing and Thinking about the Atmosphere," *Annual Review of Energy and the Environment* 27(November 2002): 1–20; Mitchell Waldrop, "An Inquiry into the State of the Earth," *Science*, 5 October 1984, 33–35; Henry Lambright, "The Ups and Downs of Mission to Planet Earth," *Public Administration Review* 54, no. 2 (March/April 1994): 99; Hans Mark to James Beggs, "Unispace 82," 25 March 1982, NASA History Office, Earth Observing System Collection, Washington, DC).

12. Richard Goody, chair, *Global Change: Impacts on Habitability—A Scientific Basis for Assessment* (Pasadena, CA: Jet Propulsion Laboratory, 1982); John McElroy and Ray Williamson, "The Evolution of Earth Science Research from Space: NASA's Earth Observing System," in *Exploring the Unknown—Selected Documents in the History of the U.S. Civil Space Program*, Volume 6: *Space and Earth Sciences*, ed. John Logsdon (Washington, DC: NASA History Series, 2004), 441–473.

13. Herbert Gursky, "Herbert Friedman," *Physics Today* 54, no. 3 (March 2001): 94.

14. Herbert Friedman, chair, *Toward an International Geosphere Biosphere Program: A Study of Global Change* (Washington, DC: National Academy Press, 1983).

15. Waldrop, "An Inquiry into the State of the Earth," 34, quotation on 35.

16. Friedman, *Toward an International Geosphere Biosphere Program*, quotation on 1; Waldrop, "An Inquiry into the State of the Earth," quotation on 34.

17. Anonymous participant quoted in Edward Edelson, "Laying the Foundation," *Mosaic* 19, no. 314 (Fall/Winter 1988): 6–7; Waldrop, "An Inquiry into the State of the Earth," 33–35.

18. Burton Edelson, "Memorandum on Global Habitability," 24 June 1983, NASA History Office, Earth Observing System Collection, Washington, DC; anonymous official, quoted in David Kennedy, "The U.S. Government and Global Environmental Change Research: Ideas and Agendas," Case C16-92-1121.0, John F. Kennedy School of Government, Harvard University (1992), 5.

19. Howard McCurdy, *The Space Station Decision: Incremental Politics and Technological Choice* (Baltimore, MD: Johns Hopkins University Press, 1990), 50; "A Talk with Burt Edelson," *Planetary Report* 2, no. 5 (September/October 1982): 17; Gruen, *The Port Unknown*, quotation on 66.

20. Mitchell Waldrop, "NASA Wants a Space Station," *Science*, 10 September 1982, 1018–1021; McCurdy, *The Space Station Decision*, 88–89, quotation on 50.

21. Burton Edelson to Pitt Thome, "Earth Remote Sensing Satellite System," 31 August 1982, NASA History Office, Earth Observing System Collection, Washington, DC.

22. "System-Z: Report of the Conceptual Study Group" and "System-Z Briefing," NASA History Office, Earth Observing System Collection, Washington, DC.

23. "System-Z: Report of the Conceptual Study Group" and "System-Z Briefing"; Waldrop, "NASA Wants a Space Station," 1020; McCurdy, 155, 169–70, 176, and 231–232.

24. Gary Taubes, "Earth Scientists Look NASA's Gift Horse in the Mouth," *Science*, 12 February 1993, 912; "System-Z Briefing."

25. Interview with unnamed former NASA official, 2005; "System-Z, Mission Analysis and Preliminary Definition Study—Research and Technology Objective and Plans Document, 146–90," Goddard Space Flight Center, NASA History Office, Earth Observing System Collection, Washington, DC.

26. Interview with unnamed former NASA official, 2005; "Staff Notes, Vol. 23, No. 40," 6 October 1988, National Center for Atmospheric Research, Boulder, CO.

27. Interview with unnamed former NASA official, 2005; Edelson, "Laying the Foundation," 7, Bretherton quotation on 7; Francis Bretherton, chair, *Earth System Science: Overview—A Program for Global Change* (Washington, DC: National Aeronautics and Space Administration, 1986), NASA History Office, Earth Observing System Collection, Washington, DC.

28. Rebecca Wright, oral history interview with Shelby Tilford, 23 June 2009, Washington, DC, tinyurl.com/y6ugp279; Burton Edelson, "Memorandum on Global Habitability."

29. "System-Z Study Project Status," 19 January 1984, NASA History Office, Earth Observing System Collection, Washington, DC.

30. Dixon Butler to Shelby Tilford, "Science and Mission Requirements Working Group for System-Z," 15 June 1983, NASA History Office, Earth Observing System Collection, Washington, DC.

31. Burt Edelson received a great deal of research material and fables about Eos, which is available at the NASA History Office, Earth Observing System Collection, Washington, DC.

32. Office of Space Science and Applications, "Use and Needs of the Space Station Complex for Science and Applications," May 1984, NASA History Office, Earth Observing System Collection, Washington, DC.

33. National Research Council, Space Studies Board, *Space Science in the Twenty-First Century, Imperatives for the Decades 1995–2015: Mission to Planet Earth* (Washington, DC: National Academy Press, 1988), quotation on ix.

34. EOS Science Group, "Eos Mission and Requirements Working Group Report—NASA Technical Memorandum 86129," August 1984, 1–10, https://ntrs.nasa.gov/archive/nasa/casi.ntrs.nasa.gov/19840022381.pdf.

35. EOS Science Group, "Eos Mission and Requirements Working Group Report," 13–62.

36. EOS Science Group, "Eos Mission and Requirements Working Group Report," 24–26.

37. Goddard Space Flight Center, EOS Project Office, *Earth Observation System: Preliminary Operations Concept and Information System Requirements* (November 1984), 2–1, NASA History Office, Earth Observing System Collection, Washington, DC.

38. Goddard Space Flight Center, EOS Project Office, *Earth Observation System.*

39. Goddard Space Flight Center, EOS Project Office, *Earth Observation System,* chapter 3.

40. John H. McElroy to Burton I. Edelson, "Preliminary Thoughts on the Utility of the Space Station to Operational Earth Observations," May 10, 1984, NASA History Office, Earth Observing System Collection, Washington, DC.

41. "Status Report: NOAA/NASA-EOS Study on the Joint Utilization of the Space Station Polar Platform," presentation to Shelby Tilford and John McElroy, 5 April 1985, NASA History Office, Earth Observing System Collection, Washington, DC.

42. "Earth Observing System: Status Review," 20 September 1985, NASA History Office, Earth Observing System Collection, Washington, DC; "NOAA/NASA-EOS Plans for Joint Utilization of Space Station Polar Platforms: Presentation of Recommendations and Actions for Next Steps," 4 December 1985, NASA History Office, Earth Observing System Collection, Washington, DC.

43. "Earth Observing System: Status Review" and "NOAA/NASA-EOS Plans for Joint Utilization of Space Station Polar Platforms."

44. "Minutes of Eos Retreat: Bethesda Marriott–Pooks Hill," 14–16 January 1986, NASA History Office, Earth Observing System Collection, Washington, DC.

45. "Minutes of Eos Retreat"; John McElroy quoted in Jay Lowndes, "NOAA Seeks Man-Tended Free Flyer," *Aviation Week and Space Technology*, 10 September 1984, 151.

## Chapter 3. A Long Road to a New Mission

1. John Logsdon, "The Evolution of Space Policy and Plans," in *Exploring the Unknown—Selected Documents in the History of the U.S. Civil Space Program*, Volume 1: *Organizing for Exploration*, ed. John Logsdon et al. (Washington, DC: NASA History Series, 1995), 392; Ronald Reagan, Executive Order 12490 (12 October 1984).

2. Thomas O'Toole, "NASA Chief Takes Leave to Fight Fraud Charges," *Washington Post*, 5 December 1985, A3.

3. Ian Fisher, "James Fletcher, 72, NASA Chief Who Urged Shuttle Pro-

gram, Dies," *New York Times*, 24 December 1991; William Rogers, chair, "Report of the Presidential Commission on the Space Shuttle Challenger Accident" (6 June 1986), https://www.govinfo.gov/content/pkg/GPO-CRPT-99hrpt1016/pdf/CHRG-101shrg1087-1.pdf.

4. Thomas Paine, chair, *Pioneering the Space Frontier: An Exciting Vision for Our Next Fifty Years in Space* (New York: Bantam Books, 1986), 1; Michael Isikoff, "NASA Chief Hails Push for Mars Colony," *Aviation Week and Space Technology*, 24 May 1986, A3; Robert Cowen, "Study on US Space Goals: Bold Vision or Space Cadet's Wish List?," *Christian Science Monitor*, 3 July 1986, 5.

5. Paine, *Pioneering the Space Frontier*, 32–33.

6. John Noble Wilford, "Threat to Nation's Lead in Space Is Seen in Lack of Guiding Policy," *New York Times*, 30 December 1986, A1; Francis Clines, "Publication Gives Edge in Space to Soviet Union," *New York Times*, 17 June 1986, C5; Craig Covault, "Fletcher Cites Turf Battles in Space Program Decision Delays," *Aviation Week and Space Technology*, 15 September 1986, 77.

7. Robert Cowen, "NASA Urges Intense Earth Surveillance," *Christian Science Monitor*, 30 June 1986, 3; Edward Edelson, "Laying the Foundation," *Mosaic* (Fall/Winter 1988): 7; Mitchell Waldrop, "Washington Embraced Global Earth Sciences," *Science*, 5 September 1986, 1040; Francis Bretherton, chair, *Earth System Science Overview: A Program for Global Change* (Washington, DC: Earth System Science Committee, NASA Advisory Council, 1986), preface.

8. Bretherton, *Earth System Science Overview*, 22–46, quotation on 22.

9. EOS Science Steering Committee, "From Pattern to Process: The Strategy of the Earth Observing System," Summer 1986, full committee report and eight more detailed panel reports available at NASA History Office, Washington, DC; memorandum, Shelby Tilford to EOS Staff, "Plans for the EOS AO and Facility Instruments Decisions," 6 June 1985, NASA History Office, Earth Observing System Collection, Washington, DC; memorandum, Shelby Tilford to EOS Staff, "Earth Observing System Facility Instruments Decisions and Plans," 13 April 1987, NASA History Office, Earth Observing System Collection, Washington, DC.

10. John A. Eddy, chair, *Global Change in the Geosphere-Biosphere: Initial Priorities for an IGBP* (Washington, DC: National Academy Press, 1986), quotations on xi, 1, and 3.

11. Eddy, *Global Change in the Geosphere-Biosphere*, 1–91, quotation on 12; David Kennedy, "The U.S. Government and Global Environmental Change Research: Ideas and Agendas," Case C16-92-1121.0, John F. Kennedy School of Government, Harvard University (1992).

12. Joseph Farman, Brian Gardiner, and Jonathan Shanklin, "Large Losses of Total Ozone in Antarctica Reveal ClOx/NOx Interaction," *Nature*, 16 May 1985, 207–210; Henry Lambright, "NASA, Ozone, and Policy Relevant Science," *Research Policy*, 1 September 1995, 752.

13. Henry Lambright, "NASA and the Environment: The Case of Ozone" (Washington, DC: NASA SP-2005-4538, 2005); *Senate Committee on the Environment and Public Works, "Ozone Depletion, the Greenhouse Effect, and Climate Change,"* S. Hrg 99-723 (10–11 June 1986), Gelman Library, Government Documents Holdings, George Washington University, Washington, DC.

14. Lambright, "NASA and the Environment"; *Senate Committee on the Environment and Public Works, "Ozone Depletion, the Greenhouse Effect, and Climate Change."*

15. Lambright, "NASA and the Environment"; *Senate Committee on the Environment and Public Works, "Ozone Depletion, the Greenhouse Effect, and Climate Change."*

16. Lambright, "NASA and the Environment"; *Senate Committee on the Environment and Public Works, "Ozone Depletion, the Greenhouse Effect, and Climate Change."*

17. "Swifter Warming of Globe Foreseen," *New York Times*, 11 June 1986, A17; "Aide Sees Need to Head Off Global Warming," *New York Times*, 12 June 1986, B8; Patrick Leahy to NASA Administrator James Fletcher, 23 June 1986 NASA History Office, Earth Observing System Collection; Roger Pielke, "Policy History of the U.S. Global Change Research Program, Part I: Administrative Development," *Global Environmental Change* (April 2000): 17.

18. Interview with Joseph K. Alexander, October 2006; anonymous aide quoted in Roger Pielke Jr., "Completing the Circle: Global Change Science and Usable Policy Information" (PhD diss., University of Colorado, 1994), 140.

19. Kennedy, "The U.S. Government and Global Environmental Change Research," 8, 9; Carol Butler, "Oral History Transcript: Anthony J. Calio," interviewed in McLean, VA, 12 April 2000.

20. Kennedy, "The U.S. Government and Global Environmental Change Research," 10.

21. Kennedy, 11.

22. Dallas Peck, chair, Committee on Earth Sciences, *Our Changing Planet: A U.S. Strategy for Global Change Research* (Washington, DC: White House Federal Coordinating Council for Science, Engineering, and Technology, 1988), appendix A.

23. Kennedy, "The U.S. Government and Global Environmental Change Research," 12.

24. NSC-NSDD-254, U.S. Space Launch Strategy (Fact Sheet), 27 December 1986, https://aerospace.csis.org/wp-content/uploads/2019/02/NSDD-254-US-Space-Launch-Strategy.pdf.

25. Charles MacKenzie, Goddard EOS project manager, to Office of Space Science and Applications, "Minimum a Budget Profile," 8 January 1987, NASA History Office, Earth Observing System Collection, Washington, DC.

26. Shelby Tilford to Thomas N. Pyke, 15 May 1987, NASA History Office, Earth Observing System Collection, Washington, DC.

27. "Outlook for Space Science: Conversations with Burt Edelson and Lennard Fisk," *Space World*, November 1987.

28. MacKenzie.

29. "Briefing to the Administrator: EOS and Polar Platforms," 8 July 1987, NASA History Office, Earth Observing System Collection, Washington, DC.

30. "Briefing to the Administrator."

31. Sally K. Ride, interview for the NASA Johnson Space Center Oral History Project, December 6, 2002, http://www.jsc.nasa.gov/history/oral_histories/a-b.htm; Sally Ride, "Leadership and America's Future in Space: A Report to the Administrator," National Aeronautics and Space Administration, August 1987, 6.

32. Ride, "Leadership and America's Future in Space," 7–25, quotations on 7, 23, and 25.

33. Ride, 25; quotation from interview with anonymous Ride Report staff member, April 2005.

34. David H. Moore, "The NASA Program in the 1990s and Beyond," Congressional Budget Office (May 1988); interview with Alexander; Robert Cowen, "NASA Hopes New Missions Will Wipe Out Old Image Problems," *Christian Science Monitor*, July 23, 1987, 6.

35. James Fletcher, Clarence Brown, and Erich Bloch to James C. Miller III, 21 October 1987, NASA History Office, files for Administrator James Fletcher, Washington, DC.

36. Kennedy, "The U.S. Government and Global Environmental Change Research," 3.

37. Kennedy, 3–5, quotations on 4 and 5.

38. Shelby Tilford to Lennard Fisk, "Reconciliation of Servicing Study Costs," 24 May 1989, NASA History Office, Earth Observing System Collection, Washington, DC.

39. Tilford to Fisk.

40. Tilford to Fisk.

41. Tilford to Fisk.

42. Lennard A. Fisk to Thomas N. Pyle Jr., 23 December 1988, NASA History Office, Earth Observing System Collection, Washington, DC.

43. Earth System Science Committee, NASA Advisory Council, "Earth System Science, A Closer View: A Program for Global Change," January 1988, NASA History Office, Earth Observing System Collection, Washington, DC.

44. National Research Council, Space Studies Board, "Space Science in the Twenty-First Century, Imperatives for the Decades 1995–2015: Mission to Planet Earth," 1988, ix and iii.

45. National Research Council, Space Studies Board, 16–51, quotations on xi and 15.

46. Michael Weisskopf, "Scientist Says Greenhouse Effect Is Setting In," *Washington Post*, 24 June 1988, A4; Michael Weisskopf, "Greenhouse Effect Fueling

Policy Makers," *Washington Post*, 15 August 1988, A1; Phillip Shabecoff, "Global Warming Has Begun, Expert Tells Senate," *New York Times*, 24 June 1988.

47. *Senate Committee on Commerce, Science, and Transportation, Hearing on Global Change Research*, S. Hrg. 100-816 (13 July 1988).

48. *Our Changing Planet*; Kennedy, "The U.S. Government and Global Environmental Change Research," 4–5.

49. Edward S. Goldstein, *NASA's Earth Science Program: A Case Study* (PhD diss., George Washington University, 2006), 146.

50. John Kingdon, *Agendas, Alternatives, and Public Policies* (New York: HarperCollins, 1995), 168; George Bush, "Remarks for the Space Shuttle Challenger Dedication," 21 March 1987, Bush Presidential Records, George Bush Presidential Library, College Station, TX.

51. George H. W. Bush, "Excerpts of Remarks at George C. Marshall Space Flight Center," 29 October 1987, Bush Presidential Records, George Bush Presidential Library, College Station, TX; Statement of Vice President George Bush on space, 19 September 1988, NASA History Office, George H. W. Bush Presidential collection, Washington, DC.

52. Bill Peterson, "Bush Vows to Fight Pollution, Install Conservation Ethic; Speech Distances Candidate from Reagan," *Washington Post*, 1 September 1988, A1; George Will, "Who's the Real Environmentalist?," *Washington Post*, 18 September 1988, C7; Goldstein, *NASA's Earth Science Program*, 150; Michael Weisskopf, "Visit with President-Elect Cheers Environmentalists," *Washington Post*, 1 December 1988.

53. Kennedy, "The U.S. Government and Global Environmental Change Research," 7–8.

54. Pielke, "Completing the Circle," 133; *Senate Committee on Commerce, Science, and Technology, Hearing on the National Global Change Research Act of 1989*, S. Hrg. 101-32 (22 February 1989); *Senate Committee on Commerce, Science, and Transportation, Hearing on the Mission to Planet Earth*, S. Hrg. 101-106 (8 March 1989).

55. *Senate Committee on Commerce, Science, and Transportation, Hearing on the Mission to Planet Earth.*

56. NASA Administrator James C. Fletcher to President George H. W. Bush, March 9, 1989, NASA History Office, Chronological Files of Administrator James C. Fletcher, Washington, DC.

57. Thor Hogan, *Mars Wars: The Rise and Fall of the Space Exploration Initiative* (Washington, DC: National Aeronautics and Space Administration, 2007), 37–76.

58. Hogan, 37–76, *Mars Wars.*

59. Kathy Sawyer, "Bush Taps Truly to Head NASA: Former Astronaut Popular on the Hill," *Washington Post*, 13 April 1989; Warren Leary, "Bush Chooses Former Astronaut to Head NASA, in a First," *New York Times*, 13 April 1989.

60. Hogan, *Mars Wars*, 37–76; Frank Martin, interviewed in Howard McCurdy, *The Decision to Send Humans Back to the Moon and on to Mars* (unpublished manuscript), NASA History Office, Washington, DC; Craig Covault, "Manned Lunar Base, Mars Initiative Raised in Secret White House Review," *Aviation Week and Space Technology*, 17 July 1989, 24.

61. *Our Changing Planet*, B2.

62. "New Earth Observing Platforms to Study Global Water, Biology," *Aviation Week and Space Technology*, 13 March 1989, 46; Shelby Tilford to EOS Staff, "Current Status of EOS," 13 July 1989, NASA History Office, Earth Observing System Collection, Washington, DC.

63. Shelby Tilford to EOS Staff; "EOS Non-Advocacy Review: Final Report of the Management Subcommittee," May 1989, NASA History Office, Earth Observing System Collection, Washington, DC; Shelby Tilford to Goddard Space Flight Center, Director of Flight Projects, "Response to the EOS Nonadvocacy Review (NAR) Committee," 20 September 1989, NASA History Office, Earth Observing System Collection, Washington, DC.

64. George H. W. Bush, "Remarks on the 20th Anniversary of the Apollo 11 Moon Landing," 20 July 1989, tinyurl.com/y9rr9v26; Richard Truly press conference transcript, 20 July 1989, NASA History Office, George H. W. Bush Presidential Collection, Washington, DC.

65. Allan Gold, "Bush Administration Is Divided on Move to Halt Global Warming," *New York Times*, 27 October 1989; Phillip Shabecoff, "White House Says Bush Will Call Meeting about Global Warming," *New York Times*, 10 May 1989; Kathy Sawyer, "Bush Urges Commitment to New Space Exploration," *Washington Post*, 21 July 1989, A4.

66. Pielke, "Completing the Circle," 123–134; *Senate Committee on Foreign Relations, Subcommittee on International Economic Policy, Trade, Oceans, and Environment, Hearing on the Paris Economic Summit and the International Environmental Agenda*, S. Hrg. 101-420 (3 August 1989).

67. Committee on Earth Sciences; NASA Fiscal Year 1990 Budget Estimates, library of the Office of the Chief Financial Officer, NASA Headquarters, Washington, DC.

68. Lennard Fisk, "Earth Observing System: Briefing to the Office of Management and Budget," September 1989, NASA History Office, Earth Observing System Collection, Washington, DC.

69. NASA Fiscal Year 1991 Budget Estimates, library of the Office of the Chief Financial Officer, NASA Headquarters, Washington, DC.

70. NASA Fiscal Year 1991 Budget Estimates.

71. David Kennedy, "This Far and No Further: The Rise and Fall of the Committee on Earth and Environmental Sciences," Case C16-92-1122.0, John F. Kennedy School of Government, Harvard University (1992), 11.

72. Kennedy, "This Far and No Further," 9.

73. NASA Fiscal Year 1991 Budget Estimates.

74. Robert Grady to Richard Truly, 6 October 1989, NASA History Office, Earth Observing System Collection, Washington, DC.

75. Richard Truly to Robert Grady, 16 October 1989, NASA History Office, Earth Observing System Collection, Washington, DC; Goldstein, *NASA's Earth Science Program*, 150.

76. Richard Truly to Robert Grady, 31 October 1989, NASA History Office, Earth Observing System Collection, Washington, DC; John Perry, "Global Change: From Rhetoric to Reality," *Reviews of Geophysics*, Supplement (April 1991): 40.

## Chapter 4. Rearguard Action to Save EOS

1. Brian Burrough, *Dragonfly: NASA and the Crisis Aboard Mir* (New York: HarperCollins, 1998), 240; Thor Hogan, *Mars Wars: The Rise and Fall of the Space Exploration Initiative* (Washington: NASA History Series, 2007), 93.

2. Allan Bromley to Frank Press, 29 January 1990, NASA History Office, Earth Observing System Collection, Washington, DC.

3. National Research Council, *The U.S. Global Change Research Program: An Assessment of 1991 Plans* (Washington, DC: National Academies Press, 1990), 1.

4. Committee on Global Change, "The U.S. Global Change Research Program: An Assessment of 1991 Plans," National Research Council, 1990, 1–63, quotations on 1, 2, 61 and 63; interview with Joseph Alexander, October 2006.

5. Committee on Global Change, "The U.S. Global Change Research Program," 70 and 62; Bob Davis, "Science Panel Backs NASA Satellite Plan, but Urges Launches of Smaller Satellites," *Wall Street Journal*, 4 April 1990, B4.

6. Bob Davis, "U.S. Climate Satellites Weather Criticism," *Wall Street Journal*, 22 March 1990; William K. Stevens, "Huge Space Platforms Seen as Distorting Studies of Earth," *New York Times*, 19 June 1990.

7. Williams; Davis, "U.S. Climate Satellites Weather Criticism."

8. Mark Albrecht to Fred McClure, 16 April 1990, Bush Presidential Records, George Bush Presidential Library; Jim Cicconi to George H. W. Bush, 30 April 1990, Bush Presidential Records, George Bush Presidential Library, College Station, TX.

9. Talking Points for President George H. W. Bush, Congressional Leadership Meeting on Space, 27 April 1990, Bush Presidential Records, George Bush Presidential Library, College Station, TX.

10. "Leadership Meeting with the President and Vice President on the Civil Space Program," memorandum from Acting Assistant Administrator for Legislative Affairs to NASA Administrator, 3 May 1990, NASA History Office, Earth Observing System Collection, Washington, DC; interview with anonymous senior congressional aide via e-mail, 15 December 2004.

11. U.S. House of Representatives, Committee on Appropriations, Department of Veterans Affairs and Housing and Urban Development, and Independent Agencies, *Appropriations for 1991*, Part 4: *National Aeronautics and Space Administration* (Washington, DC: U.S. Government Printing Office, 1990); *US Senate, Committee on Appropriations, Departments of Veterans Affairs and Housing and Urban Development, and Independent Agencies, Appropriations, FY91, Part 2*, S. Hrg. 101-1099 (Washington, DC: U.S. Government Printing Office, 1990).

12. Warren Leary, "Hubble Telescope Loses Large Part of Optical Ability: Most Complex Instrument in Space Is Crippled by Flaw in a Mirror," *New York Times*, 27 June 1990, A1; Bob Davis, "NASA Finds Hubble Mirror Is Defective," *Wall Street Journal*, 28 June 1990.

13. Joyce Price, "Chief Calls NASA Funding 'Crucial for U.S. Survival," *Washington Times*, 3 July 1990; quotation in John Burgess, "Can U.S. Get Things Right Anymore? Hubble Telescope, Space Shuttle Problems Raise Questions about American Technology," *Washington Post*, 3 July 1990.

14. Dan Quayle, *Standing Firm* (New York: Harper Collins, 1994), 184; Hogan, *Mars Wars*, 122–123.

15. Quayle, *Standing Firm*, 184; Hogan, *Mars Wars*, 122–123.

16. Timothy Curry and Lynn Shibut, "The Cost of the Savings and Loan Crisis: Truth and Consequences," *FDIC Banking Review* (December 2000): 26–35, http://www.workingre.com/wp-content/uploads/2013/08/cost-of-SL.pdf; Dan Balz, "The Power of 'Me'—With Word Added by Democrats, Bush Erased His Pledge on Taxes," *Washington Post*, 1 July 1990, A1.

17. Hogan, *Mars Wars*, 125; Public Law 101-611, National Aeronautics and Space Administration Authorization Act, Fiscal Year 1991.

18. *Report of the Advisory Committee on the Future of the U.S. Space Program* (Washington, DC: Government Printing Office, 1990), 2–4.

19. Report of the Advisory Committee on the Future of the U.S. Space Program, 5.

20. Report of the Advisory Committee on the Future of the U.S. Space Program, 5; Memorandum, Mark Albrecht to Dan Quayle, 7 December 1990, Bush Presidential Records, George Bush Presidential Library, College Station, TX.

21. Hogan, *Mars Wars*, quotations on 127 and 128; Mark Carreau, "Panel Wants to Phase Out Space Shuttle: White House Backs Changes That Will Transform NASA," *Houston Chronicle*, 11 December 1990, 1.

22. Public Law 101–606, *U.S. Global Change Research Act of 1990*, November 16, 1990; David Kennedy, "This Far and No Further: The Rise and Fall of the Committee on Earth and Environmental Sciences," Case C16-92-1122.0, John F. Kennedy School of Government, Harvard University (1992), 12–13

23. Kennedy, 14–15, quotations on 15.

24. Kennedy, 16; Committee on Earth and Environmental Sciences, "Our Changing Planet: The FY 1992 Global Change Research Program," 1991, tinyurl.com/ycv6a3zv; U.S. House of Representatives, Committee on Appropriations, Department of Veterans Affairs and Housing and Urban Development, and Independent Agencies, *Appropriations for 1992*, Part 6: *National Aeronautics and Space Administration House of Representatives* (Washington, DC: U.S. Government Printing Office, 1991).

25. U.S. House of Representatives, Committee on Appropriations, Department of Veterans Affairs and Housing and Urban Development, and Independent Agencies, *Appropriations for 1992*, Part 6: *National Aeronautics and Space Administration House of Representatives.*

26. U.S. House of Representatives, Committee on Appropriations, Department of Veterans Affairs and Housing and Urban Development, and Independent Agencies, *Appropriations for 1992*, Part 6: *National Aeronautics and Space Administration.*

27. *US Senate, Committee on Appropriations, Departments of Veterans Affairs and Housing and Urban Development, and Independent Agencies, Appropriations, FY92*, Part 2, S. Hrg. 102-113 (Washington, DC: U.S. Government Printing Office, 1991).

28. Earth Observing System External Engineering Review, Terms of Reference, March 1991; NASA Management Instruction NMI 1152.69, "Earth Observing System (EOS) Engineering Review Advisory Committee," 8 May 1991, NASA History Office, Earth Observing System Collection, Washington, DC.

29. Quotations in Erik Conway, *Change in the Air: A History of Atmospheric Sciences at NASA*, unpublished manuscript; "Latest EOS Reviewers Given Free Reign to Suggest Changes," *Space News*, 13 May 1991, 18.

30. "Latest EOS Reviewers Given Free Reign to Suggest Changes."

31. Quotation in U.S. House of Representatives, Departments of Veterans Affairs and Housing and Urban Development, and Independent Agencies, Appropriations Bill, H.R. Rep. 102-94 (3 June 1991); "House Approves Rescue of Space Station," *New York Times*, 7 June 1991, D16; Jill Graham, "Space Station Saved after Pleas from Walker, Other NASA Supporters," State News Service, 6 June 1991; William J. Broad, "House Vote Sets Stage for Conflict between Two Allies in Space Program," *New York Times*, 8 June 1991, A7.

32. Broad, "House Vote Sets Stage for Conflict," A7; Kathy Sawyer, "Space Budget Battle: Humans 1, Robots 0," *Washington Post*, 19 June 1991, A17.

33. Sawyer, "Space Budget Battle," A7; U.S. Senate, Departments of Veterans Affairs and Housing and Urban Development, and Independent Agencies, Appropriations Bill, S. Rep. 102-107 (11 July 1991).

34. U.S. Senate, Departments of Veterans Affairs and Housing and Urban Development, and Independent Agencies, Appropriations Bill, S. Rep. 102-107 (11 July 1991).

35. Douglas Isbell, "EOS Panel Favors Shrinking Satellites," *Space News*, 29 July 1991, 4; James Hansen, William Rossow, and Inez Fung, "The Missing Data on Global Climate Change," *Issues in Science and Technology* 7, no. 1 (Fall 1990): 62–69; Hogan, *Mars Wars*, 95; Earth Observing System Engineering Review Committee, Meeting Minutes, 18–26 July 1991, provided to the authors by a senior NASA official, February 2007.

36. Earth Observing System Engineering Review Committee, Meeting Minutes; *US House of Representatives, Committee on Science, Space, and Technology, Hearings on the Earth Observing System Engineering Review*, 26 September 1991; quotation in *US Senate, Committee on Commerce, Science, and Transportation, Hearing on the NASA Earth Observing System*, S. Hrg. 102-647, 26 February 1992.

37. Earth Observing System Engineering Review Committee, Meeting Minutes; quotation in *US Senate, Committee on Commerce, Science, and Transportation, Hearing on the NASA Earth Observing System.*

38. *Report of the Earth Observing System Engineering Review Committee*, September 1991, NASA History Office, Earth Observing System Collection, Washington, DC.

39. *US House of Representatives, Committee on Science, Space, and Technology, Hearings on the Earth Observing System Engineering Review*, 26 September 1991.

40. U.S. House of Representatives, Making Appropriations for the Departments of Veterans Affairs and Housing and Urban Development, and for Sundry Independent Agencies, Commissions, Corporations, and Offices for Fiscal Year Ending September 30, 1992, and for Other Purposes, H.R. Rep. 102-226 (27 September 1991); Edward S. Goldstein, *NASA's Earth Science Program: A Case Study* (PhD diss., George Washington University, 2006), 185; *National Aeronautics and Space Administration FY 1993 Budget Request to Congress, Congressional Back-up Data Book*, library of NASA's Chief Financial Officer, Washington, DC.

41. Richard H. Truly, administrator, National Aeronautics and Space Administration, "Letter of Request for Review of NASA's EOSDIS Plans" to Dr. Frank Press, president, National Academy of Sciences, December 1991, NASA History Office, Earth Observing System Collection, Washington, DC.

42. Lennard Fisk, statements before the U.S. Senate Committee on Commerce, Science, and Transportation.

43. *US Senate, Committee on Commerce, Science, and Transportation, Hearing on the NASA Earth Observing System*; National Aeronautics and Space Administration, *Report to Congress on the Restructuring of the Earth Observing System*, 9 March 1992, NASA History Office, Earth Observing System Collection, Washington, DC.

44. National Aeronautics and Space Administration, *Report to Congress on the Restructuring of the Earth Observing System.*

45. National Aeronautics and Space Administration, *Report to Congress on the Restructuring of the Earth Observing System.*

46. *US Senate, Committee on Commerce, Science, and Transportation, Hearing on the NASA Earth Observing System.*

47. National Research Council, "Panel to Review EOSDIS Plans: Final Report," 11 January 1994, tinyurl.com/yatxtsad.

48. National Research Council, "Panel to Review EOSDIS Plans: Final Report."

49. National Research Council, "Panel to Review EOSDIS Plans: Final Report."

## Chapter 5. Last Hard Push to Secure EOS

1. Bryan Burroughs, *Dragonfly: NASA and the Crisis Aboard Mir* (New York: HarperCollins Publishers, 1998), 239–243.

2. Thor Hogan, *Mars Wars: The Rise and Fall of the Space Exploration Initiative* (Washington: NASA History Series, 2007), 53.

3. Kathy Sawyer, "Truly Fired as NASA Chief, Apparently at Quayle's Behest: Ex-Astronaut Feuded with Space Council," *Washington Post*, 13 February 1992, A1; William J. Broad, "NASA Chief Quits in Policy Conflict," *New York Times*, 13 February 1992, A1; Craig Covault, "White House to Restructure Space Program: Truly Fired," *Aviation Week and Space Technology*, 17 February 1992, 18.

4. Henry Lambright, *Transforming Government: Dan Goldin and the Remaking of NASA* (Arlington, VA: PricewaterhouseCoopers Endowment for the Business of Government, March 2001), 15.

5. Lambright, *Transforming Government*; Burroughs, *Dragonfly*, 244.

6. Lambright, *Transforming Government*, 14–15.

7. *US Senate, Committee on Commerce, Science, and Transportation, Nomination of Daniel S. Goldin to Be Administrator of the National Aeronautics and Space Administration*, S. Hrg. 102-744 (27 March 1992); Lambright, *Transforming Government*, 14–15.

8. Lambright, *Transforming Government*, 14–15.

9. Charles F. Bolden, letter to Officials-in-Charge, "Review of NASA," 7 May 1992, NASA History Office, Earth Observing System Collection, Washington, DC.

10. Craig Covault, "Goldin Orders Sweeping Review of NASA Programs, Eyes 30% Cuts," *Aviation Week and Space Technology*, 1 June 1992.

11. Covault, "Goldin Orders Sweeping Review of NASA Programs"; "Goldin to Announce Budget Review Teams; 'Realistic' Options Due June 24," *Aerospace Daily*, 26 May 1992, 307; "Goldin: No Sacred Cows in Search for Cuts to Fund New Technology," *Aerospace Daily*, 27 May 1992, 314.

12. William Townsend to Lennard Fisk, 21 June 1992, NASA History Office, Earth Observing System Collection, Washington, DC.

13. Townsend to Fisk; National Aeronautics and Space Administration FY 1994 Budget Request to Congress, Congressional Back-up Data Book, library of the NASA Office of the Chief Financial Officer, Washington, DC; National Aeronau-

tics and Space Administration, *1995 MTPE EOS Reference Handbook*, 18–21, NASA History Office, Earth Observing System Collection, Washington, DC.

14. Townsend to Fisk; National Aeronautics and Space Administration FY 1994 Budget Request to Congress; National Aeronautics and Space Administration, *1995 MTPE EOS Reference Handbook.*

15. National Aeronautics and Space Administration, "Adapting the Earth Observing System to the Projected $8 Billion Budget: Recommendations from the EOS Investigators," 14 October 1992.

16. U.S. House of Representatives, Making Appropriations for the Departments of Veterans Affairs and Housing and Urban Development, and for Sundry Independent Agencies, Commissions, Corporations, and Offices for Fiscal Year Ending September 30, 1993, and for Other Purposes, H.R. Rep. 102-902 (24 September 1992).

17. Lambright, *Transforming Government*, 16.

18. Lambright, 17.

19. Major Garret, "Deeper Cuts May Be Near; The White House Signals Flexibility," *Washington Times*, 4 March 1993; Eric Pianin, "House and Senate Negotiators Near Agreement on Budget Plan," *Washington Post*, 31 March 1993, A7; Burroughs, *Dragonfly*, 262–264.

20. U.S. House of Representatives, Departments of Veterans Affairs and Housing and Urban Development, and Independent Agencies Appropriations Bill, 1994, H.R. Rep. 103-150 (22 June 1993); Leon Panetta to Dan Goldin, "Budget Realities," 18 June 1993, NASA History Office, Earth Observing System Collection, Washington, DC; Jack Fellows to Dan Goldin, "FY 1995 Budget Process," 15 July 1993, NASA History Office, Earth Observing System Collection, Washington, DC; interview with senior NASA official, March 2007.

21. U.S. House of Representatives, Making Appropriations for the Departments of Veterans Affairs and Housing and Urban Development, and for Sundry Independent Agencies, Commissions, Corporations, and Offices for Fiscal Year Ending September 30, 1993, and for Other Purposes, H.R. Rep. 102-902 (24 September 1992); U.S. House of Representatives, Making Appropriations for the Departments of Veterans Affairs and Housing and Urban Development, and for Sundry Independent Agencies, Commissions, Corporations, and Offices for Fiscal Year Ending September 30, 1994, and for Other Purposes, H.R. Rep. 103-273 (4 October 1993).

22. Panetta to Goldin; interview with senior NASA official.

23. James Asker, "Earth Mission Faces Growing Pains," *Aviation Week and Space Technology*, 21 February 1994, 36.

24. Interview with Charles F. Kennel, NASA Johnson Space Center Oral History Project, 21 October 2002, NASA History Office, Washington, DC.

25. Interview with Charles F. Kennel.

26. National Aeronautics and Space Administration, *1995 MTPE/EOS Reference*

*Handbook*; Greg Williams, *The Difficult Journey of a Great Idea: An Inside History of Earth System Science at NASA*, December 2005, NASA History Office, Earth Observing System Collection, Washington, DC.

27. Williams, *The Difficult Journey of a Great Idea*; Berrien Moore, "Responding to the Demand to Reduce the Earth Observing System Budget from $8 to $7.25 Billion: 1990–2000," *Earth Observer* (July/August 1994): 4; National Aeronautics and Space Administration FY 1995 Budget Request to Congress, Congressional Back-up Data Book, library of the NASA Office of the Chief Financial Officer, Washington, DC.

28. Henry Lambright, "Downsizing Big Science: Strategic Choices," *Public Administration Review* 58, no. 3 (May/June 1998): 259.

29. Andrew Lawler, "NASA Mission Gets Down to Earth," *Science*, September 1, 1995, 1208.

30. U.S. Senate, Departments of Veterans Affairs and Housing and Urban Development, and Independent Agencies, Appropriations Bill, S. Rep. 103-137 (9 September 1993).

31. George E. Brown, chair, Committee on Science, Space, and Technology, U.S. House of Representatives, to D. James Baker, Acting Undersecretary for Oceans and Atmospheres, U.S. Department of Commerce, February 1993, NASA History Office, Earth System Science Collection, Washington, DC.

32. Jim Exon, chair, Subcommittee on Nuclear Deterrence, Arms Control and Defense Intelligence, U.S. Senate, to Ron Brown, Secretary of Commerce, 3 June 1993, in *Exploring the Unknown—Selected Documents in the History of the U.S. Civil Space Program*, Volume 3: *Using Space*, ed. John M. Logsdon et al. (Washington, DC: NASA History Office, 1998).

33. Bill Clinton and Al Gore, *Putting People First* (New York: Time Books, 1992), 23–24.

34. Department of Commerce, "Establish a Single Civilian Operational Environmental Polar Satellite Program," September 1993, in *Exploring the Unknown*, Volume 3; White House, "Presidential Decision Directive NSTC-2: Convergence of U.S. Polar-Orbiting Operational Environmental Satellite Systems," 5 May 1994, in *Exploring the Unknown*, Volume 3; Pamela Mack and Ray Williamson, "Observing the Earth from Space," in *Exploring the Unknown*, Volume 3.

35. Roger Pielke Jr., "Completing the Circle: Global Change Science and Usable Policy Information" (PhD diss. University of Colorado, 1994), 221–231; Henry Lambright, "The Rise and Fall of Interagency Cooperation: The U.S. Global Change Research Program," *Public Administration Review* 57, no. 1 (January/February 1997): 41.

36. Andrew Lawler, "Walker Unveils R&D Strategy," *Science*, 23 December 1994, 1938; Andrew Lawler, "Science and Technology Policy Headed for Political Maelstrom," *Science*, 18 November 1994, 1152; "House Panel May Target Mission

to Planet Earth," *Space News*, 19 December 1994, quotation on 3; Andrew Lawler, "Robert Walker: The Speaker's Right Hand on Science," *Science*, 11 August 1995, quotation on 794.

37. Lambright "Downsizing Big Science," quotation on 264; Report on NASA's Zero Base Review in National Aeronautics and Space Administration FY 1997 Budget Request to Congress, Congressional Back-up Data Book, library of the NASA Office of the Chief Financial Officer, Washington, DC.

38. Presidential Review Directive NSTC-1, "Interagency Review of Federal Laboratories," 5 May 1994.

39. NASA Advisory Council, *NASA Federal Laboratory Review* (Washington, DC: U.S. Government Printing Office, 1995), 41–45, quotation on 45.

40. Williams, *The Difficult Journey of a Great Idea*.

41. Interview with former White House official, March 2007.

42. Quotation in National Aeronautics and Space Administration, "Understanding Our Changing Planet: NASA's Mission to Planet Earth," Fact Book, May 1995, NASA History Office, Earth Observing System Collection, Washington, DC; Williams, *The Difficult Journey of a Great Idea*; Michael King, "Editor's Corner," *Earth Observer* (May/June 1995), tinyurl.com/ydd4nfh2.

43. Robert Walker to Dan Goldin, 6 April 1995; Robert Walker to Bruce Alberts, 6 April 1995; interview with Charles F. Kennel.

44. U.S. House of Representatives, Setting Forth the Congressional Budget for the United States Government of the Fiscal Years 1996, 1997, 1998, 1999, 2000, 2001, and 2002, H.R. Rep. 104-120 (15 May 1995); Edward S. Goldstein, *NASA's Earth Science Program: A Case Study* (PhD diss., George Washington University, 2006), quotation on 200.

45. U.S. House of Representatives, Committee on Appropriations, Department of Veterans Affairs and Housing and Urban Development, and Independent Agencies, *Appropriations for 1996*, Part 7: *National Aeronautics and Space Administration House of Representatives* (Washington, DC: U.S. Government Printing Office, 1995).

46. U.S. House of Representatives, Committee on Science, Subcommittee on Space and Aeronautics, *Fiscal Year 1996 NASA Authorization* (Washington, DC: U.S. Government Printing Office, 1995).

47. Kathy Sawyer, "Republican Budget Proposals Put Cost-Cutting Initiatives at Risk," *Washington Post*, 15 May 1995, quotation on A17; quotation in Daniel Goldin, "Hands Off Mission to Planet Earth," *Space News*, 1995.

48. U.S. General Accounting Office, "NASA Earth Observing System: Estimated Funding Requirements," Report to the Chairman, Committee of Science, House of Representatives, June 1995; *US Senate, Committee on Commerce, Science, and Transportation, NASA Mission to Planet Earth Program*, S. Hrg. 104-242 (Washington, DC: U.S. Government Printing Office, 1995); U.S. House of Rep-

resentatives, Departments of Veterans Affairs and Housing and Urban Development, and Independent Agencies, Appropriations Bill, H.R. Rep. 104-201 (21 July 1995); Lawler, "NASA Mission Gets Down to Earth," quotation on 1208; Warren Ferster, "New EOS Plan Targets $12 Billion in Savings," *Space News*, 31 July 1995, quotation on 4.

49. Charles Kennel, "MTPE/EOS Introduction," presentation to National Research Council Committee on Global Change Research, 19–28 July 1995, NASA History Office, Earth Observing System Collection, Washington, DC; Robert Harriss, "MTPE/EOS Science Strategy," presentation to National Research Council Committee on Global Change Research, 19–28 July 1995, NASA History Office, Earth Observing System Collection, Washington, DC; William Townsend, "MTPE/EOS Program Overview," presentation to National Research Council Committee on Global Change Research, 19–28 July 1995, NASA History Office, Earth Observing System Collection, Washington, DC; Robert Winokur, "NOAA/NASA Roundtable Discussions," presentation to National Research Council Committee on Global Change Research, 19–28 July 1995, NASA History Office, Earth Observing System Collection, Washington, DC; Christopher Scolese, "EOS Flight Segment," presentation to National Research Council Committee on Global Change Research, 19–28 July 1995, NASA History Office, Earth Observing System Collection, Washington, DC; National Aeronautics and Space Administration, *1999 EOS Reference Handbook: A Guide to NASA's Earth Science Enterprise and the Earth Observing System*, 1999; National Aeronautics and Space Administration, FY 1994 Budget Request to Congress, Congressional Back-up Data Book, library of the NASA Office of the Chief Financial Officer, Washington, DC.

50. Kennel, "MTPE/EOS Introduction"; Harriss, "MTPE/EOS Science Strategy"; Townsend, "MTPE/EOS Program Overview"; Winokur, "NOAA/NASA Roundtable Discussions"; Scolese, "EOS Flight Segment"; National Aeronautics and Space Administration, *1999 EOS Reference Handbook*; National Aeronautics and Space Administration, FY 1994 Budget Request to Congress.

51. Kennel, "MTPE/EOS Introduction"; Harriss, "MTPE/EOS Science Strategy"; Townsend, "MTPE/EOS Program Overview"; Winokur, "NOAA/NASA Roundtable Discussions"; Scolese, "EOS Flight Segment"; National Aeronautics and Space Administration, *1999 EOS Reference Handbook*; National Aeronautics and Space Administration, FY 1994 Budget Request to Congress.

52. Kennel, "MTPE/EOS Introduction"; Harriss, "MTPE/EOS Science Strategy"; Townsend, "MTPE/EOS Program Overview"; Winokur, "NOAA/NASA Roundtable Discussions"; Scolese, "EOS Flight Segment"; National Aeronautics and Space Administration, *1999 EOS Reference Handbook*; National Aeronautics and Space Administration, FY 1994 Budget Request to Congress.

53. Kennel, "MTPE/EOS Introduction"; Harriss, "MTPE/EOS Science Strategy"; Townsend, "MTPE/EOS Program Overview"; Winokur, "NOAA/NASA

Roundtable Discussions"; Scolese, "EOS Flight Segment"; National Aeronautics and Space Administration, *1999 EOS Reference Handbook*; National Aeronautics and Space Administration, FY 1994 Budget Request to Congress.

54. Goldstein, 185 {{KCJ: Provide full citation information}}.

55. National Research Council, *A Review of the U.S. Global Change Research Program and NASA's Mission to Planet Earth/Earth Observing System* (Washington, DC: National Academies Press, 1995), 1–25, quotations on 23 and 24–25.

56. U.S. Senate, Departments of Veterans Affairs and Housing and Urban Development, and Independent Agencies, Appropriations Bill," S. Rep. 104-140 (13 September 1995).

57. Andrew Lawler, "Walker Sets Off Alarm Bells with Efforts to Rein in EOS," *Science*, 16 February 1996, quotations on 900; U.S. House of Representatives, Making Appropriations for the Departments of Veterans Affairs and Housing and Urban Development, and for Sundry Independent Agencies, Commissions, Corporations, and Offices for Fiscal Year Ending September 30, 1996, and for Other Purposes, H.R. Rep. 104-140 (6 December 1995).

58. Reshaping Options Implementation Study, "Presentation to the Administrator," 12 February 1996.

59. U.S. House of Representatives, Making Appropriations for the Departments of Veterans Affairs and Housing and Urban Development, and for Sundry Independent Agencies, Commissions, Corporations, and Offices for Fiscal Year Ending September 30, 1997, and for Other Purposes, H.R. Rep. 104-812 (20 September 1996).

## Conclusion

1. John W. Kingdon, *Agendas, Alternatives, and Public Policies* (New York: HarperCollins College Publishers, 1995), 71–196.

2. Kingdon, *Agendas, Alternatives, and Public Policies.*

3. Anthony Downs, *Inside Bureaucracy* (Boston, MA: Little, Brown and Company, 1966), quotation on 10; James Q. Wilson, *Bureaucracy: What Government Agencies Do and Why They Do It* (New York: Basic Books, 1989), 188–189.

4. National Aeronautics and Space Act of 1958, Public Law 85-568, 29 July 1958, tinyurl.com/y9cr39j8; Richard LeRoy Chapman, *A Case Study of the U.S. Weather Satellite Program: The Interaction of Science and Politics* (PhD diss., Syracuse University, 1967), 297–300.

5. Wilson, *Bureaucracy*, 190; Pamela E. Mack and Ray A. Williamson, "Observing the Earth from Space," in *Exploring the Unknown—Selected Documents in the History of the U.S. Civil Space Program*, Volume 3: *Using Space*, ed. John Logsdon et al. (Washington, DC: Government Printing Office, 1998), 155–177; Chapman, *A Case Study of the U.S. Weather Satellite Program*, 117–121.

6. Downs, *Inside Bureaucracy*, 13–14.

7. *Senate Committee on the Environment and Public Works, Hearing before the Subcommittee on Environmental Pollution, "Ozone Depletion, the Greenhouse Effect, and Climate Change,"* S. Hrg. 99-723 (10–11 June 1986); "A Dire Forecast for Greenhouse Earth," *Washington Post*, 11 June 1986, A1; "Swifter Warming of Globe Foreseen," *New York Times*, 11 June 1986, A17; "Greenhouse Effect Needs More Study, U.S. Aide Says," *Washington Post*, 12 June 1986, A17; "Aide Sees Need to Head Off Global Warming," *New York Times*, 12 June 1986, B8.

8. Roger Pielke Jr., "Completing the Circle: Global Change Science and Usable Policy Information" (PhD diss., University of Colorado, 1994), quotation on 140; David Kennedy, "The U.S. Government and Global Environmental Change Research: Ideas and Agendas," Case C16-92-1121.0, John F. Kennedy School of Government, Harvard University, quotation on 10.

9. *Report of the Advisory Committee on the Future of the U.S. Space Program* (Washington, DC: Government Printing Office, 1990), 2–21, quotations on 2 and 20–21.

10. Howard McCurdy, "Organizational Decline: NASA and the Life Cycle of Bureaus," *Public Administration Review* 51, no. 4 (July-August 1991): 163, 174, and 310–311.

11. Downs, *Inside Bureaucracy*, 10; *Report of the Advisory Committee on the Future of the U.S. Space Program*, 13.

12. Dan Quayle, *Standing Firm* (New York: HarperCollins, 1994), quotation on 180; Bryan Burroughs, *Dragonfly: NASA and the Crisis Aboard Mir* (New York: HarperCollins Publishers, 1998), quotation on 244.

13. Interview with Lennard Fisk, November 2006; Thor Hogan, *Mars Wars: The Rise and Fall of the Space Exploration Initiative* (Washington: NASA History Series, 2007), 159–166; Quayle, *Standing Firm*, quotations on 185–186.

14. Wilson, *Bureaucracy*, 203–204.

15. Wilson, 203–204; interview with Joseph K. Alexander, October 2006.

16. Quayle, 184; Edward Goldstein, *NASA's Earth Science Program: A Case Study* (PhD diss., George Washington University, 2006), 183.

17. John W. Kingdon, *Agendas, Alternatives, and Public Policies* (New York: HarperCollins College Publishers, 1995), 105.

18. Interview with anonymous Office of Management and Budget official, March 2006.

19. Ellen Gray and Patrick Lynch, "Earth from Space: 15 Amazing Things in 15 Years," National Aeronautics and Space Administration, 18 December 2014, tinyurl.com/yd6qosld.

20. Gray and Lynch, "Earth from Space."

21. Quotation in Gray and Lynch; Umair Irfan, "Greenland's Ice Sheet Is Driving Global Sea Level Rise," *Vox*, 14 December 2017, tinyurl.com/yck6alr7.

22. Gray and Lynch, "Earth from Space."

23. Gray and Lynch.

24. Committee on the Decadal Survey for Earth Science and Applications from Space, *Thriving on Our Changing Planet: A Decadal Strategy for Earth Observation from Space* (Washington, DC: National Academies Press, 2018).

25. Committee on the Decadal Survey for Earth Science and Applications from Space, S-1 and 1-2.

26. Committee on the Decadal Survey for Earth Science and Applications from Space, 2-23.

27. National Academies of Sciences, Engineering, and Medicine, *Accomplishments of the U.S. Global Change Research Program* (Washington: National Academies Press, 2017), 11–16, quotations on 11, 13, and 16.

28. National Academies of Sciences, Engineering, and Medicine, 16–23, quotations on 16 and 22.

# Index

*Note: page number in italics refer to figures.*


ABMA. *See* Army Ballistic Missile Agency
Advanced Research Projects Agency (ARPA), 31
Advanced Spaceborne Thermal Emission and Reflection Radiometer (ASTER), 146
Advisory Committee on the Future of the U.S. Space Program. *See* Augustine Committee
aerial photography. *See* remote sensing by aircraft or satellite
aeronautics research in NASA
  Beggs' emphasis on, 52
  and FY1992 budget cuts, 138
aerosols and clouds, EOS data on, 202
aerospace industry, and space policy agenda, 21
*Agendas and Instability in American Politics* (Baumgartner and Jones), 21
Agriculture and Resource inventory Surveys through Aerospace Remote Sensing (AgRISTARS), 40
Air Force, U.S., research on remote sensing by satellite, 30
AIRS. *See* Atmospheric Infrared Sounder
Albrecht, Mark
  on Augustine Report, 129
  career of, 108
  and EOS Engineering Review Committee, 137
  and NASA leadership, 149, 150
  and NASA mission planning, 108–109
  and NASA reform, 125, 193
  and SEI program, 119–120
  as Space Council Executive Secretary, 108
  and Truly, firing of, 149
American Society of Photogrammetry, 29
Apollo program, 13
Applications Technology Satellite (ATS) program, 33, 41
Army Air Service, aerial mapping surveys, 29
Army Ballistic Missile Agency (ABMA), 8–9, 31
ARPA. *See* Advanced Research Projects Agency
ASTER. *See* Advanced Spaceborne Thermal Emission and Reflection Radiometer
Atmospheric Infrared Sounder (AIRS), 146, 154
ATS. *See* Applications Technology Satellite (ATS) program
Augustine, Norm, 126, 127, 132–135
Augustine Committee (Advisory Committee on the Future of the U.S. Space Program), 125–126
Augustine Report, 26, 127–130, 132–135, 191–192

Baker, D. James, 121–122
Baumgartner, Frank. *See* Comparative Agendas Project (CAP); punctuated equilibrium model of policy change (Baumgartner and Jones)
Beggs, James
  career of, 51
  departure as NASA administrator, 77
  Edelson and, 54
  and global change research at NASA, 61, 198–199
  and NASA budget cuts, 53–54
  priorities for NASA, 52
  selection as NASA administrator, 51, 52, 86–87
  and space station, 17, 52, 63, 65, 69, 198
  and Truly, firing of, 149
Bloch, Erich, 88, 96
BOB. *See* White House Bureau of the Budget
Bretherton, Francis
  and global change research at NASA, 66–67, 80–81
  and NRC Committee on Global Change Research, 172
  on NRC plan for earth science research, 104
  Senate testimony on climate change, 102

Bretherton Reports, 26, 80–81, 83, 85, 88, 91, 99–100, 180, 199
Bromley, D. Allan, 115, 120, 130–131, 194
Brown, George, 137, 163
bureaucracies
autonomy, external and internal elements of search for, 195
constraints on expansion of, 24
natural tendency to expand, 23
tactics for fashioning core constituency, 23–24
*See also* life cycle of bureaucracies
*Bureaucracy* (Wilson), 194
Bush, George H. W.
and Augustine Report, 191–192
breaking of no-new-taxes pledge, 126
budget cuts of FY1992, 131–132
climate change policy, criticisms of, 113
deficit reduction as priority for, 115
election of, as policy window for MTPE/EOS adoption, 181–182
and election of 1992, 157, 187
and Goldin as NASA administrator, 151
interest in environmental issues, 105–106, 183, 187, 199
interest in space policy, 104–105
space program policy, announcement of, 111–112
space program policy, reactions to, 112–113
State of the Union mentions of environment, *186*, 187
support for global change research, 106
support for manned space flight, 105, 109, 112
support for space program, 17, 20
and Truly, firing of, 149
and U.S. Global Change Research Program, 1–2, 182
Bush, George W., 200
Butler, Dixon, 68, 144

Calio, Anthony J., 62, 86–87, 88–89, 191
CAP. *See* Comparative Agendas Project
carbon cycle, EOS data on, 201–202, 205
Carter, James E. "Jimmy," 52, 53
CEES. *See* Committee on Earth and Environmental Sciences
CES. *See* Committee on Earth Sciences
Chafee, John H., 83, 85, 190
*Challenger* space shuttle, loss of
Bush's meeting with families, 105
and change in Shuttle use policy, 89–90
impact on EOS program, 89–90, 180–181, 199
impact on space program, 76, 77, 78–79
investigation of, 78
as policy window for MTPE/EOS adoption, 181
*Chandra* (X-Ray Observatory), 15–16
chlorofluorocarbons, banning of, 84, 85–86
CIAP. *See* Climatic Impact Assessment Program
Clarke, Arthur C., 13
client bureaucracy, NASA as, 194–197
climate change
articles per year in New York Times (1981–2000), 183, *184*
EOS data on, 201–202, 204–206
Gallup Polls on public view of importance (1981–2000), 183, *185*
politicians' delay in responding to, 190–191
refocusing of EOS program toward, 141
scientific community's delay in demanding action on, 181
Senate hearings on, 83–85, 101–103, 106–107
U.S. failure to act in 1980s, 84–86, 101–103
Watson Senate testimony on, 84
*See also* global change/earth science research at NASA
Climatic Impact Assessment Program (CIAP), 46–47
Clinton, William J. "Bill"
deficit reduction as priority for, 157–158
and government shutdowns, 176
and interagency cooperation in global change research, 165
NASA budget cuts under, 158–159, 161–163, 167–169, 171–178
and National Space Council, 157
program to improve government efficiency, 163–164, 166
space program as low priority for, 157
State of the Union mentions of environment, *186*, 187
support for environmental issues, 187
clouds and aerosols, EOS data on, 202
Cohen, Aaron, 119–120
Cold War space race, 6–13
and first human in space, 11–12
and manned mission to moon, 12–13
and NASA, founding of, 10–11
and space station, 52, 79
*Sputnik* and, 7, 9, 10
and U.S. spending on space exploration, 9–10

Committee for an International Geosphere-Biosphere Program, 82
Committee on Earth and Environmental Sciences (CEES)
Clinton administration disbanding of, 164
establishment of, 130
resistance to FY1992 budget cuts, 131
Committee on Earth Sciences (CES)
and coordination of global change research, 96–97, 107
and distinction of focused vs. contributing research, 110
establishment of, 88, 191
exemplary coordination management by, 115
function of, as issue, 88–89
and FY1990 budget, 113, 118
and FY1991 budget, 115
and global change research under Bush, 106
and MTPE planning, 109–110
renaming of, 130
report on earth science budgets (*Our Changing Planet*), 26, 103–104
working group of, 96–97
Committee on Environment and National Resources, 165
Committee on Global Change
and Committee on Earth Sciences, 97
establishment of, 82
and restructuring of MTPE/EOS program, 171–172, 175–176
review of USGCRP and EOS, 120–121
Communications Satellite Corporation, 13
Comparative Agendas Project (CAP), 182–188, *184–186*
*Compton* Gamma Ray Observatory, 15–16
Comsat Corporation, 53, 54
Congress
and Bush space agenda, 112–113
and FY1992 budget, 143
hearings on climate change, by year (1981–2000), 184–187, *186*
and SEI program, lack of support for, 123–124
and space policy agenda, 20–21
support for MTPE, 124
Congressional Budget Office (CBO)
on cost of EOS program, 95
on cost of MTPE, 107
Corell, Robert, 97, 106–107, 115, 131–132
Council of Scientific Unions, 107
craft agencies, 195

Darman, Richard, 108–109
Defense Meteorological Satellite Program (DMSP), 33
Department of Agriculture, U.S. (USDA), and satellite remote sensing development, 40
Department of Defense (DOD)
and early rocket research, 8
proposal to merge NOAA and DOD weather satellite programs, 163–164
and space shuttle program, 52
DMSP. *See* Defense Meterological Satellite Program
Domestic Policy Council, 191
Downs, Anthony, 18, 22–23, 188. *See also* life cycle of bureaucracies
Dresler, Paul, 103–104, 106, 115

*Early Bird* satellites, 13
Earth Observation System (EOS). *See* EOS
Earth Resources Observation Satellite (EROS) program, 37–38
Earth's atmosphere
evolution of, 45
interaction with biologic actors, 45–46
earth system science. *See* global change/earth science research
*Earth System Science: A Closer View* (Second Bretherton Report), 26, 99–100
Earth System Science Committee (ESSC)
Bretherton Reports on global change research, 26, 80–81, 83, 85, 88, 91, 99–100, 180, 199
establishment of, 67
as first to equate earth system science and global change research, 80
second Bretherton Report (*Earth System Science*), 26, 99–100
Earth System Science Pathfinder Program, 173, 174, 177
Eddy, John "Jack," 82, 121
Eddy Report, 82, 83, 121, 180, 199
Edelson, Burton
career of, 54
and global change research at NASA, 59, 60, 62, 66–67, 76, 198
and MTPE program, 100
naming of EOS, 68
and NASA as aging bureaucracy, 190
and NASA-NOAA relations, 54–55
and OSSA, 54, 67

Edelson, Burton (*continued*)
retirement from NASA, 91
and space science agenda, 69, 72
and space station program, 62, 63, 69, 72
and System-Z development, 63–64, 65, 66, 198
Eisenhower, Dwight D., 8, 9, 10–11, 31, 35
election of 1960, 11–12
election of 1988, 105
election of 1994, 165
environmental movement, supersonic transport (SST) and, 46–48
Environmental Research Institute (University of Michigan), 35–36
EOS (Earth Observation System)
announcement of opportunity to science community, 99
Augustine Report criticisms of, 129, 133–134
breadth and complexity of, as challenge, 71–72
budget cuts, effects of, 196–197, 200–201
and budgets of 1990s, 114–118, 127, 131–136, 139–140, 143, 144, 155–157, 159, 162–163, 177
and *Challenger* loss, impact of, 89–90, 180–181, 199
Clinton plan to merge NOAA weather satellite program with, 163–164
Commission on Space endorsement of, 79
components of, 118
cost projections for, 95, 98, 114, 118, 140, 145, 171
cost reductions, NASA's search for, 90, 111
data, valuable contributions of, 200–206
data processing requirements for, 70–71, 75
delays in launching
budget cuts and, 135–136, 177
and critics' calls to scale down satellites, 122
NOAA concerns about, 90–91
and NOAA withdrawal from program, 99
reviews of program and, 132
delays in program development, 74–76, 90
dependence on manned space flight as issue, 76, 90
development of objectives and requirements for, 69–71
Engineering Review Committee study of, 26, 136–138, 140–143, 147–148, 155, 196
first budget appropriation for, 1, 118, 119
Fisk's efforts to restart, 91
and focused vs. contributing research types, 111
funding of, as issue, 75
Goldin's NASA reform program and, 154–155, 167, 200
independent assessment, changes recommended by, 110–111
instruments aboard
challenges to plans for, 121–122, 133, 134–135, 137, 140–143
Goldin cost-cutting reforms and, 154
independent assessment of, 110–111
NASA selection of, 110
original plans for EOS-A and -B, 114, 121–122
restructuring under Clinton budget cuts, 172–174
scaling back, for smaller satellites, 145–146
scaling back, to maintain political support, 162–163
launching of satellites
issues surrounding, 74, 91–92
launch platform as issue in, 91–92, 97–98, 99
number launched so far, 3, 200
revisions of plans for, 141–142
launch of first satellite (*EOS-AM I*), 178
maintenance costs as issue in, 75
multiple agencies participating in, 114
and NOAA sensing projects, 72–76, 90–91, 99, 168, 169, 172–173
NRC review of program, 120–122
OMB call for review of, 132, 194
personnel needed for management of, 71
planned life of, 74
reconfiguration around smaller satellites, 142, 143–146, 199–200
and launch order, debate on, 146
satellites and payloads in, 145–146
scaling back of science goals for, 143, 145, 157, 159
and scientific community as constraint on NASA autonomy, 195–197
refocusing toward climate change, 141, 145
renaming of System-Z as, 68, 199
research teams, selection of, 110
restructuring under Clinton budget cuts, 161–163, 167–169, 171–178
and abandonment of original EOS concept, 174
and changes to science missions, 172–174, 175

establishment of biennial review process, 174
and partnerships with industry, 173
Ride Report recommendations on, 94–95
scientific community and, 122–123, 195–197
servicing of, issues surrounding, 92–93, 98–99
shuttle payload limitations and, 91–92
and simultaneity, 91, 120, 121–122, 137, 141
stepped implementation plans for, 74
technological advances required to implement, 94
*See also* MTPE/EOS; System-Z
EOS Data and Information System (EOSDIS)
and Clinton budget cuts, 161–162
cost projections for, 145
EOS Engineering Review Committee on, 142–143, 147–148, 155
EOS Science Group report on, 81–82
and FY1992 budget, 143
and FY1993 budget, 144, 155–156
Goldin's NASA reform program and, 154–155
NSO review recommendations on, 174–175
Reshaping Options Implementation Study on, 177
restructuring of, 147–148, 155
Zraker Committee review of, 144, 175
EOSDIS. *See* EOS Data and Information System
EOS Engineering Review Committee, 26, 136–138, 140–143, 147–148, 155, 196
EOS Program Office, 91, 146, 155, 158. *See also* EOS Project Office
EOS Project Office, 68, 71–73, 74, 75, 98–99, 110. *See also* EOS Program Office
EOS Science Group, 69–70, 71, 73, 75, 81–82
EROS. *See* Earth Resources Observation Satellite (EROS) program
ESA. *See* European Space Agency
ESSC. *See* Earth System Science Committee
European Space Agency (ESA), and EOS, 74
Executive Office of the President, and space policy agenda, 19–20
"Exploration of Space with Rocket Devices" (Tsiolkovsky), 4
*Explorer I* satellite, 9, 10, 14, 30

Federal Aviation Administration (FAA), 46
Federal Coordinating Council for Science, Engineering, and Technology, 115. *See also* Committee on Earth Sciences (CES)
Fellows, Jack, 88, 95, 103, 131–132, 141
First Symposium on Remote Sensing of the Environment, 35–36
Fisk, Lennard
and Augustine Report House hearings, 134–135
career of, 91
and EOS program, 91, 98, 99, 110–111, 144–145, 146
and Hubble Space Telescope malfunction, 124
on MTPE, 114
as NASA Chief Scientist, 156
on National Space Council influence, 193, 194
and Senate hearings on MTPE, 106–108
Fletcher, James
appointment as NASA administrator, 78
and Bush, lobbying of, 108
career of, 78
and climate science research, 47–49
and Commission on Space report, 78
and EOS program, 90, 92–93, 97–98
and global change research at NASA, 86
and interagency cooperation on global change research, 95–96
and NASA priorities, development of, 79–80
and Office of Exploration, creation of, 95
and space shuttle program, 47, 48
and Truly, firing of, 149
forest fires, EOS data on, 202
Friedman, Herbert, 26, 59–61, 67, 197
Friedman Report (*Toward an International Geosphere-Biosphere Program*), 26, 60–61, 180, 200
Frieman, Edward
and EOS Engineering Review Committee study, 136–138, 141–142, 143, 180, 196
and MPTE associate administrator post, 160
and NRC Committee on Global Change Research, 172
Frieman Committee. *See* EOS Engineering Review Committee
*From Pattern to Process* report (EOS Science Group), 81

Gagarin, Yuri, 12
*Gaia* (Lovelock), 45
*Galileo*, 15, 53
Gallup Polls, on public view of climate change importance (1981–2000), 183, *185*
garbage can theory (Cohen, March, and Olsen), 18

General Accounting Office (GAO), report on MTPE/EOS, 171
Geostationary Operational Environmental Satellite (GOES), 35
global change/earth science research
  Bush's emphasis on, 106, 182
  concerns about cost of, 87–88
  EOS retreat from, for political reasons, 162–163
  focused vs. contributing research types in, 109–110, 116
  and FY1990 budget, 113–114
  and FY1991 budget, 114–118
  and NASA agenda, 2
  questionable support for in mid-1980s, 83
  struggle for control of, 86–89
global change/earth science research at NASA
  Clinton budget cuts and, 162
  Commission on Space endorsement of, 79
  creation of Office of Space Science and Applications, 67
  data system requirements for, 70–71
  development, overview of, 1–3
  development of agenda for, 55–61
  funding as issue in, 56–57, 66
  and interagency cooperation, calls for, 94–95, 101, 103–104
  and international cooperation, calls for, 94–95, 101
  leadership as issue in, 57
  linking to space shuttle program, 198
  linking to space station program, 69, 81, 180, 198, 200
  missions on agenda in early 1980s, 67–68
  NRC proposals for, 104
  reasons for introducing, 197–198
  reasons for pursuing despite setbacks, 198–199
  reports on post-*Challenger* direction for, 80–83
  and scientific community, 57, 61–62, 194–197, 198
  second Bretherton Committee Report on, 99–100
  Space Studies Board report on, 100–101
  turn of NASA satellite programs to, 42–43, 47–50
  *See also* climate change; EOS (Earth Observation System)
*Global Change: Impacts on Habitability* report, 25–26
Global Change Research Act of 1990, 102, 103, 106, 113, 130, 164
Global Habitability initiative, 61–62
Global Habitability Report. *See* Goody Report
Global Learning and Observations to Benefit the Environment (GLOBE), 177
global perspective on earth
  development of concept, 43–45
  development of NASA research program based on, 43–45
global warming. *See* climate change
GLOBE. *See* Global Learning and Observations to Benefit the Environment
Goddard, Robert Hutchings, 4–5
Goddard Space Flight Center (GSFC)
  and early remote sensing research, 37
  and EOS program, 78, 93
  and Nimbus program, 83
  and ozone hole research, 83
  and space station program, 62
  and System-Z development, 65, 66
  *See also* EOS Project Office
GOES. *See* Geostationary Operational Environmental Satellite
Goldin, Daniel
  appointment as NASA administrator, 151, 192
  career of, 150
  and EOS, criticism of, 150–151
  and EOS redesign under Clinton budget cuts, 166, 167–169, 171, 174–175, 176
  and Kennel as MPTE associate administrator, 160–161
  and NASA budget cuts under Clinton, 158
  NASA reform program of, 150, 151–155, 156–157, 167, 200
  and NOAA/NASA collaboration, calls for, 169
  retention by Clinton administration, 157
  and review of NASA field centers' efficiency, 167
  and Walker's challenge to MTPE/EOS, 169–170, 171
Goody, Richard, 57–59, 197
Goody Report, 58–59, 61, 67, 180, 198
Gore, Al
  and Clinton space policy, 157, 161, 163–164
  criticism of National Space Council, 150, 151
  and National Performance Review, 163–164
  reorganization of White House science policymaking process, 164–165
Grady, Robert, 108, 111, 118
Graham, William
  as acting NASA administrator, 77

and global change research at NASA, 87, 88, 96
as presidential science adviser, 87, 191
at Senate hearings on climate change, 83–84, 85
greenhouse effect
EOS data on, 201–202
first direct observation in atmosphere, 41–42
Hansen congressional testimony on, 101
GSFC. *See* Goddard Space Flight Center

Hall, Mike, 96–97, 131–132
Halley's Comet, NASA mission to study, 53
Hansen, James, 84, 85, 87, 101, 141
Harriss, Bob, 161, 162, 169–170, 176
High Resolution Imaging Spectrometer, 154
Hollings, Fritz, 102, 106, 107
House appropriations committee for NASA
and EOS budget, Walker's efforts to cut, 170–171
and FY1992 budget, 137
and FY1994 budget, 158
and FY1996 budget, 171
hearings on Augustine Report, 132–135
Republican leadership after 1994 election, 165
House Committee on Science, Space, and Technology, 113, 137
House Subcommittee on Space and Aeronautics, 170–171
House VA-HUD Appropriations Subcommittee, 176–177
Hubble Space Telescope, 15, 124

IGY. *See* International Geophysical Year
Improved TIROS Operational Satellite (ITOS), 35
Infrared Interferometer Spectrometers (IRIS), 35, 41–42
interagency cooperation in global change research
calls for, 94–95, 101, 103–104
Clinton administration dismantling of, 165
decay of, in early 1990s, 130
obstacles to, 103–104
intergovernmental Panel on Climate Change (IPCC), and EOS program, 141
International Geophysical Year (IGY), U.S. plan to launch satellite for, 6–7, 8
International Geosphere Biosphere Program, and MTPE, 107
IRIS. *See* Infrared Interferometer Spectrometers
ITOS. *See* Improved TIROS Operational Satellite

Janus program, 31
Japan, and EOS, 84
Jet Propulsion Laboratory (JPL), 8, 11, 14, 43, 45, 62, 66
Johnson, Lyndon B., 10, 11, 12, 16, 38
Johnson, Richard, 86, 88
Johnson Space Center (JSC)
and planning of human exploration program, 109
and SEI implementation, 119–120
Jones, Bryan. *See* Comparative Agendas Project (CAP); punctuated equilibrium model of policy change (Baumgartner and Jones)
JSC. *See* Johnson Space Center (JSC)
Jupiter rocket program, 8–9

Keeling, Charles, 42, 180
Kennedy, David, 97, 115, 130
Kennedy, John F., 11–13
Kennel, Charles
appointment as MPTE associate administrator, 160–161
departure from NASA, 178
and EOS redesign under Clinton budget cuts, 167–169, 171–172, 174, 177
and GAO report on MTPE/EOS, 171
and NOAA/NASA collaboration, 168, 169
and NRC Committee on Global Change Research, 172
and Walker's challenge to MTPE/EOS, 171
Kingdon, John, 104, 196. *See also* multiple streams model of policy change (Kingdon)
Korolev, Sergei, 6, 7

LACIE. *See* Large Area Crop Inventory Experiment
Lambright, Henry, 26–27, 165
Landsat program
data from, 197
expansion of, 40
first satellites launched by, 39–41
funding for, 38–39, 159–160
MTPE and, 172
and NASA softening-up of decision-makers, 180
privatization plans for, 53
value of data from, 40–41, 49–50

Large Area Crop Inventory Experiment
(LACIE), 40
Launius, Roger, 4, 8
Lawler, Andrew, 166, 176–177
*Leadership and America's Future in Space* (Ride Report), 26, 93–95, 96, 128, 181
Leahy, Patrick, 86
Lewis, Jerry, 165, 176–177
life cycle of bureaucracies
and cooperative programs, danger of, 189
and MTPE/EOS, establishment of, 22–24, 188–197
and NASA as aging bureaucracy, 189–194
Augustine Report on, 191–192
exogenous factors in, 192–194
MTPE/EOS program as effort to reverse, 190
NASA's turf battles with NOAA and, 188–189
NASA's vigor as young agency, 188
research on, 18
Limb Infrared Monitor of the Stratosphere (LIMS), 43
LIMS. *See* Limb Infrared Monitor of the Stratosphere
Logsdon, John, 16, 149
Lovelock, James, 45–46
Lujan, Manuel, 130–131
*Lunar Orbiter* program, 14

*Magellan* probe, 15
manned space flight
Bush initiative to return to, 105, 109, 112
and Cold War space race, 11–12
dependence of EOS on, as issue, 76
System-Z program as rationale for, 65
*See also* Space Exploration Initiative (SEI)
Margulis, Lynn, 45–46
*Mariner* probes, 14–15
Mark, Hans, and global change research at NASA, development of, 57–58, 197–199
career of, 51–52
departure as NASA deputy administrator, 77
as NASA deputy administrator, 52
priorities for NASA, 52
and space shuttle program, 52
and space station, 65
Mars exploration program, 14–15
and global perspective on earth, development of, 43–45
Ride Report recommendations on, 94
*Mars Wars* (Hogan), 19
McCurdy, Howard, 26, 192
McDougall, Walter, 9, 30
McElroy, John, 25, 57–58, 72, 76
media
on Bush space agenda, 112–113
coverage of Senate climate change hearings, 102
response to NASA missteps of 1990, 124–125
and space policy agenda, 21
Mercury program, 11, 12, 36
"A Method of Reaching Extreme Altitudes" (Goddard), 5
Mikulski, Barbara, 123, 124, 135, 139
Mission to Planet Earth (MPTE). *See* MTPE (Mission to Planet Earth)
moon landings, 12–13
moon outpost, Ride Report recommendations on, 94
Moore, Berrien, 161, 171–172
MSS. *See* multispectral scanners
MTPE (Mission to Planet Earth)
Augustine Report on, 128–129, 133, 134
and budgets of 1990s, 127, 133, 139, 171, 176, 177
Bush's support for, 105, 109, 112, 113, 199
and Clinton administrative reforms, 167
components of, 107, 114
Congressional support for, 124
contributions to understanding climate crisis, 1, 3
cost of, as issue, 107
drop in support, after NASA misteps of 1990, 124–125
earth probe missions in, 100, 114–115, 143, 177
efforts to develop interagency agreement on, 95–97
focused vs. contributing research types, 109–110, 111
goals of, 1–2
opposition to, 2
presidential approval of, 118
repeated redesigns, factors driving, 2–3
Reshaping Options Implementation Study on, 176, 177
restructuring under Clinton budget cuts, 171–178
Ride Report recommendations on, 94
scaling back, to maintain political support, 162–163
Senate hearings on, 106–108

supporters' concerns about NASA priorities, 113
Truly's support for, 109
MTPE/EOS
budget cuts, effects of, 2
decoupling from space shuttle program, 90–92, 180–181, 199
GAO report on, 171
and NASA autonomy, constraints on, 195–197
reports on, 25–26
separation from space station program, 89–90, 111, 180
value of policy-based focus on, 24–25, 197
Walker's challenges to, 165, 169–171, 176–177
*See also* EOS (Earth Observation System); MTPE (Mission to Planet Earth)
MTPE/EOS adoption
Comparative Agendas Project (CAP) on, 182–188, *184–186*
level of public concern about global warming and, 183–184, *185*
life cycle of bureaucracies and, 22–24, 188–197
link to global change research and, 180, 182
multiple streams model of, 180–182
NASA's ability to maintain political support for, 199–200
NASA's support from science community and, 194–197
opening of policy windows for, 181–182
primary sources on, 27
process of, 199
punctuated equilibrium model on, 22, 182–188
role of scientific community in, 21, 27–28, 199
scholarship on, 24–27
softening-up of decisionmakers and, 180–181
multiple streams model of policy change (Kingdon), 18–19, 179–180
and MTPE/EOS adoption, 180–182
on softening-up of decisionmakers, 179–180
on successful policy outcomes, key to obtaining, 19
multispectral scanners (MSS), aboard *Landsat I* satellite, 39–40

NASA
administrator, and space policy agenda, 20
Advanced Program Studies Office, 80
and ATS program, 33, 41
and Augustine Report, 125–126, 127–130, 132–134
budgets
for 1960–2000, 192, *193*
in 1970, 16
in 1980s, 15
in 1990s, 108–109, 113–117, 126–127, 131–140, 143, 155–160, 166
during Apollo program, 13
Augustine Committee on, 127–128
cuts under Reagan, 51, 53–54
in post-Apollo period, 17–18
and bureaucratic conflict, experience with, 50
Chief Scientist, creation of position, 156
and Collier Trophy, 40
constraints on autonomy of, 195–197
and early satellite remote sensing projects, 31–32, 37–38
and environment research, shift toward, 42–43, 47–50
founding of, 10–11, 188
and global change research policy, struggle to control, 86–87, 88
and human exploration of space, high cost of, 108–109
and interagency cooperation on global change research, 80–81
and Landsat program, 37–41
and National Space Council, sour relations with, 119–120, 130, 196
Office of Exploration, creation of, 95
Office of Space Science and Applications, 14, 49
and ozone depletion research, 47–50, 83–84
planetary exploration of 1960s-80s, 15
research agenda, Space Studies Board reports on, 55–57
research outside manned space program, 13
rocket failures of mid-1980s, 79
siege mentality in Nixon administration, 16
and SMS program, 41
space science activities of twentieth century, 14–16
Stratosphere Research Program, 48
and weather satellites
development of technology, 34–35
joint projects with NOAA, 72–76, 90–91, 99, 168, 169, 172–173
turf disputes over programs, 32–33, 49, 188–189
*See also* life cycle of bureaucracies

NASA Advisory Council
and global change research, 66–67
and NASA priorities after *Challenger*, 80
review of NASA centers and programs, 166–167
National Academy of Sciences, Engineering, and Medicine, 204
National Aeronautics and Space Act of 1958, 11, 188
National Climate Act of 1978, 83
National Commission on Space
establishment of, 77
and policy windows for MTPE/EOS adoption, 181
report on future of space program (*Pioneering the Space Frontier*), 78–79
and Ride Report, 93
National Oceanographic and Atmospheric Administration (NOAA)
Clinton plan to merge with EOS program, 163–164
creation of, 33
and EOS, coordination with, 72–76, 90–91, 99, 168, 169, 172–173
and EOS funding, 75
and global change research policy, struggle to control, 86–87, 88
and interagency cooperation on global change research, 62, 80–81, 95–97
and ozone depletion research, 48
privatization plans for weather satellites, 53
proposal to merge weather satellite programs with DOD, 163–164
strained relations with NASA in early 1980s, 54–55
turf battles with NASA, 32–33, 49, 188–189
National Operational Meteorological Satellite System (NOMSS), 32, 33
National Performance Review, 163–164
National Polar-Orbiting Operational Environmental Satellite System, 164, 167, 172
National Research Council (NRC)
and global change research, development of, 59, 82, 198
plan for earth science research, 104
and USGCRP policy, 164
*See also* Committee on Global Change; Space Studies Board
National Science and Technology Council, 166
National Science Foundation (NSF)
and global change research, struggle for control of, 88, 95–97
and NASA global change research, 62, 80–81
National Security Council (NSC), and space policy agenda, 17, 19–20
National Security Decision Directive 254, 89–90
National Space Council
and Bush space agenda, 112
Clinton's disbanding of, 157
and EOS Engineering Review Committee study, 141
and FY1992 budget cuts, 132
Gore's criticisms of, 150, 151
hostility to EOS program, 137
influence on space policy, 193–194, 196
and NASA, sour relations between, 119–120, 130, 196
and NASA missteps of 1990, 125
and review of EOS program, 120–121
and space policy agenda, 20
takeover of space program policymaking, 120
NEMS. *See* Nimbus E Microwave Spectrometer
New Millennium Program, 173
*New York Times* climate change articles per year (1981–2000), 183, *184*
Nimbus E Microwave Spectrometer (NEMS), 42
Nimbus program, 32, 33, 34–35
data from, 197
and NASA/NOAA turf battles, 189
NASA's ending of, 54–55
and ozone depletion research, 83, 84
remote sensing equipment used in, 35, 41–42, 43
shift toward environment research, 43
UARS and, 49
value of data from, 41–42, 49–50
Nixon, Richard M., 10, 11–12, 16, 17, 33, 39, 78
NOAA. *See* National Oceanographic and Atmospheric Administration
NOMSS. *See* National Operational Meteorological Satellite System
non-governmental organizations, and space policy agenda, 21
North American Aviation, Rocketdyne division, 7
NRC. *See* National Research Council
NSC. *See* National Security Council

Oberth, Hermann, 4, 5
Ocean Topography Experiment (TOPEX), 67, 68, 81, 114, 132

Office of Management and Budget (OMB)
Bush's emphasis on global change research and, 106
call for review of EOS program, 132, 194
and FY1991 budget, 114–118
and FY1992 budget, 131
and global change research, funding for, 87–88, 96
and Landsat funding, 39
and NASA mission planning, 16, 17, 19–20, 53, 108–109
on OES cost, 95
and satellite privatization plans, 53
and space policy agenda, 19–20
and USGCRP funding rules, 130–131
Office of Manned Space Science (OMSS), 36–37
Office of Science and Technology Policy (OSTP), 19–20, 80–81, 86, 87
Office of Space Science and Applications (OSSA), 49, 54, 67, 75, 156
OMSS. *See* Office of Manned Space Science
Operational Satellite Improvement Program (OSIP), 33–34, 55
OSIP. *See* Operational Satellite Improvement Program
OSSA. *See* Office of Space Science and Applications
OSTP. *See* Office of Science and Technology Policy
*Our Changing Planet* report (Committee on Earth Sciences), 26, 103–104
ozone depletion research
calls for increase in, 48
and chlorofluorocarbon ban, 84, 85–86
in mid-1970s, 46–50
in mid-1980s, 83–84
NASA and, 47–50, 83–84
as policy window for MTPE/EOS adoption, 181

Paine, Thomas, 39, 77, 149
Panetta, Leon, 158
Peck, Dallas, 62, 96, 102–103, 113, 130, 131
photogrammetry, development of technology, 29
*Pioneer 4*, 14
*Pioneering the Space Frontier* (National Commission on Space), 78–79
polar ice, EOS data on, 201
policy alternatives generation
in multiple streams model, 19
in space policy, 20–21
policy change models. *See* multiple streams model of policy change (Kingdon); punctuated equilibrium model of policy change (Baumgartner and Jones)
policy monopolies, punctuated equilibrium model on, 21–22
policy windows
administration changes as, 104
Kindgon on, 19
MTPE/EOS program and, 62, 72, 83, 118, 181–182
and NASA early satellite programs, 50
Presidential Decision Directive NSTC-2, 163–164
president(s)
as space policy agenda-setter, 19–20
State of the Union Addresses, number of reference to climate change (1981–2000), 184–185, *186*, 187
process streams in organizations
garbage can theory on, 18
*See also* multiple streams model (Kingdon)
Project Vanguard, 8–9
Project WS-117L, 31
punctuated equilibrium model of policy change (Baumgartner and Jones), 18, 21–22
and MTPE/EOS adoption, 182–188
on policy monopolies, 21–22
on softening-up of decisionmakers, 179–180
Pyke, Thomas, 90, 99

Quayle, Dan
and Augustine Committee, 125–126, 132
and EOS Engineering Review Committee, 137
and EOS program, resistance to, 193, 194, 196
and NASA agenda, 108–109
and NASA as aging bureaucracy, 193
on NASA misteps of 1990, 125
and National Space Council, 20, 108, 112
and SEI program, 120, 123
and Truly, firing of, 149

RAND Corporation, 30
*Ranger* spacecraft, 14
RCA (Radio Corporation of America), 31, 37
Reagan, Ronald W.
and *Challenger* explosion, 78, 89–90
and global change research, 83, 86, 87, 190, 191
and Landsat program, 53
and NASA budget cuts, 51, 53–54
and National Commission on Space, 77

Reagan, Ronald W. (*continued*)
new NASA leadership under, 51–52
and space station, 17, 69
Redstone rocket, 8
*Remote Measurement of Pollution* (1972 study), 42–43
remote sensing by aircraft or satellite
development of technology, 29–31
role in global change research, 56, 58–59, 60
shuffling of bureaucratic control over (1950s), 31
*See also* satellite remote sensing
*Report of the Advisory Committee on the U.S. Space Program* (Augustine Report), 26, 127–130, 132–135, 191–192
Republican Congress of 1994
budget reduction efforts of, 170–171
challenges to MTPE/EOS program, 177
and government shutdowns, 176
support for privatization of applied R&D, 165–166
Reshaping Options Implementation Study, 176, 177
Ride, Sally, 26, 80, 93, 95, 199
Ride Report, 26, 93–95, 96, 128, 181

Sagan, Carl, 44, 45, 55
SAMOS. *See* Satellite Military Observation System
SAMS. *See* Stratospheric and Mesospheric Sounder
satellite communications, NASA research on, 13
Satellite Infrared Spectometers (SIRS), 35, 42
Satellite Military Observation System (SAMOS), 31
satellite remote sensing
development of civilian environmental monitoring programs, 35–39
development of technologies for, 39–42
early programs in, 31–32
first sounding instruments in space, 35
multispectral scanners and, 39–40
repeatability of measurements from, 40
research on uses of data from, 36
satellite size, early debate on, 37
shuffling of bureaucratic control over (1950s), 31
vs. aircraft-based sensors, 38, 39, 40
*See also* EOS (Earth Observation System); System-Z; weather satellite programs
Scanning Microwave Spectrometer (SCAMS), 42
Schirra, Walter, 36
scientific community
announcement of EOS opportunities to, 99
calls for coordination of efforts with NASA, 57
delays in pushing for climate change action, 181
and global change research at NASA, development of, 197
initial rejection of NASA leadership, 61–62, 198
NASA as client bureaucracy to, 194–197
role in MTPE/EOS establishment, 21, 27–28, 199
and System-Z, alliance created by, 198
SEI. *See* Space Exploration Initiative
Senate appropriations committee for NASA
and FY1992 budget, 135–136, 139–140
and FY1993 budget, 155–156
and FY1994 budget, 159
and FY1996 budget, 176
Republican leadership after 1994 election, 165
Senate Committee on Commerce, Science, and Transportation, 102, 106–108, 171
Senate Committee on Environment and National Resources, 101
Senate Committee on Outer Space, 10
Senate Committee on the Environment and Natural Resources, 101
Senate Foreign Relations Committee, 113
Senate hearings on climate change, 83–85, 101–103, 106–107
Senate Subcommittee for Science, Technology and Space, 144–145
Senate Subcommittee on Environmental Pollution, 83–85
Shea, Eileen, 88, 96
Sheppard, Alan, 12
simultaneity in EOS data
challenges to need for, 120, 121–122, 137, 141
value of, 65, 91
SIRS. *See* Satellite Infrared Spectometers
SMS. *See* Synchronous Meteorological Satellite (SMS) program
solar system exploration, Ride Report recommendations on, 94
Soviet Union, early rocket and space programs, 6. *See also* Cold War space race
Space and Earth Science Advisory Committee, 67
space exploration

early fictional fantasies about, 3–4
early rocket research, 4–6
Space Exploration Initiative (SEI)
Augustine Report on, 133
Bush announcement of, 112–113
Bush effort to save, 123–124
Clinton budget cuts and, 158, 159
cost projections for, 120
and FY1992 budget cuts, 138
loss of support for, 120, 123–124
NASA study on implementation of, 119–120
reasons for collapse of, 193
removal of funds for, 126
space policy
key agenda-setting actors in, 19–21
Ride Report recommendations on, 93–95
space program, agenda for, Augustine Committee effort to set, 127–128
space science programs at NASA
Augustine Committee on, 128
Augustine Report on, 132–134
Beggs' emphasis on, 52
budget cuts of FY1992 and, 138–149
Goldin reforms and, 156–157
space shuttle program
Augustine Committee on, 128
Carter administration and, 52
*Challenger* loss and, 76, 77, 78–79
decoupling of MTPE/EOS and, 90–92, 180–181, 199
linking of earth science research to, 198
original plan as joint NASA and DOD asset, 52
origins of, 16–17
and ozone depletion, 47
space shuttles
and EOS payload, 91–92
grounding of (1990), effect on space program, 124–125
space station
Augustine Committee on, 128
Beggs' push for development of, 52
and budget cuts of FY1992, 138–139
Bush support for, 112
Clinton budget cuts and, 158
Cold War space race and, 52, 79
defining of missions for, 62, 63
development of, 16–17
development of requirements for, 69
linking of global change/earth science research to, 69, 81, 180, 198, 200
NASA design specifications for, 63
NASA misteps of 1990 and, 125
Reagan's announcement of, 69
separation from MTPE/EOS, 89–90, 111, 180
and space science agenda, 100
and System-Z, coordination of development with, 66, 68, 69
space station, international, protection from budget cuts, 166
space station support satellites, proposals for, 62
Space Station Task Force, 62–63
Space Station Working Group, 69
Space Studies Board
and global change research at NASA, 197, 199
and NASA softening-up of decision-makers, 180
report on EOS data, 202–204
reports on NASA research goals, 26, 55–57, 67, 69, 100–101
Spin-Scan Cloud Camera (SSCC), 33, 41
*Sputnik*, 7, 9, 10
*Sputnik III*, 11
SSCC. *See* Spin-Scan Cloud Camera
state control of technology development, as product of space race, 9
State of the Union Addresses, number of reference to climate change (1981–2000), 184–185, *186*, 187
*A Strategy for Earth Science from Space in the 1980s*, Part I: *Solid Earth and Oceans* (Space Studies Board, 1982), 26, 55–57
Stratospheric and Mesospheric Sounder (SAMS), 43
Sununu, John, 125
supersonic transport (SST) project, 46–48
*Surveyor* series spacecraft, 14
Synchronous Meteorological Satellite (SMS) program, 41
Synoptic Terrain Photography Experiment, 36
System-Z
data processing requirements for, 65
and other NASA earth science initiatives, 67–68
proposal for development of, 63–65, 198–199
proposed instrumentation and capabilities, 64–65
renaming as EOS (Earth Observation System), 68, 199

System-Z (*continued*)
and scientific community, alliance with, 198
and simultaneous data, value of, 65
and space station, coordination of development with, 66, 68, 69
study of requirements for, 65–66, 68
System-Z Science Group, 68

TDRSS. *See* Tracking and Data Relay Satellite System
telecommunications, NASA research on, 13
Television Infrared Observing Satellite (TIROS) program, 31–33, 41, 42, 189
*Telstar* satellite, 13
Tikhonravov, Mikhail Klavdiyevich, 6–7
Tilford, Shelby
career of, 67, 157
and EOS Engineering Review Committee, 137
and EOS program, 90–91, 98
and EOS restructuring, 144, 146, 155, 157
and FY1992 budget cuts, 131–132
and global change research at NASA, 61, 67, 96
as MTPE associate administrator, 157
and MTPE program, 100, 109
and NOAA-NASA joint ventures, 54–55
resignation of, 157, 160
and System-Z development, 68
TIROS. *See* Television Infrared Observing Satellite (TIROS) program
TIROS Operational System (TOS), 32
TOPEX. *See* Ocean Topography Experiment
TOS. *See* TIROS Operational System
Total Ozone Mapping Spectrometer mission, 114–115, 132
*Toward an International Geosphere-Biosphere Program* (Friedman Report), 26, 60–61, 180, 200
Tracking and Data Relay Satellite System (TDRSS), 72, 74
Traxler, Robert, 123–124, 132–135, 138–139
Tropical Rainforest Mapping Mission, 115, 132
Truly, Richard
appointment as NASA administrator, 109
and Augustine Report, 125–126, 130
and Bush space agenda, 112
career of, 109
and EOS Engineering Review Committee, 137
and EOS program, 117–118, 136, 182, 187, 199
firing of, 149–150, 191–192
and manned flight program, concerns about, 109
and MTPE, support for, 109
on NASA priorities, 109
and reconfiguration of EOS around small satellites, 144
and Senate hearings on FY1992 budget, 135
White House concerns about, 119
and Zraker Committee review of EOSDIS, 144
Trump, Donald, 200

UARP. *See* Upper Atmosphere Research Program
UARS. *See* Upper Atmosphere Research Satellite
UNISPACE II (Vienna, 1982), 58, 61–62, 198
Upper Atmosphere Research Program (UARP), 49, 114, 132, 180
Upper Atmosphere Research Satellite (UARS), 49, 180
USGCRP. *See* U.S. Global Change Research Program
U.S. Geological Survey (USGS)
and interagency cooperation on global change research, 80–81
and satellite remote sensing technology, 37–39, 40
U.S. Global Change Research Program (USGCRP)
Augustine Committee on, 129
as Bush priority, 182, 199
and Clinton administrative reforms, 167
constraints on use of funding for, 130–131
and FY1992 budget cuts, 132
Gore's criticisms of, 164
and MTPE/EOS, 1–2, 114
NRC review of, 120
presidential approval of, 118
scientific community's criticisms of, 122–123
value of data from, 204–205

V-2 rockets, 5–6, 7
Van Allen, James, 6, 7, 122
Van Allen radiation belts, discover of, 9
venue access, and MTPE/EOS adoption, 184–187, *186*
Venus Orbiting Imaging Radar (VOIR), 53
*Viking* missions, 8, 14–15, 44
VOIR. *See* Venus Orbiting Imaging Radar (VOIR)
von Braun, Wernher, 5–6, 8–9
*Voyager* program, 15

Walker, Robert
as chair of House Science and Technology Committee, 165
challenges to MTPE/EOS mission and budget, 165, 169–171, 176–177
and privatization of applied R&D, 165–166
retirement from Congress, 178
support for human spaceflight program, 138
Watson, Robert, 83–85, 87, 102, 162, 165
Weather Bureau, turf disputes over weather satellite programs, 32–33, 188–189
weather satellite programs, 31–35
development of technology for, 33–35
proposal to merge NOAA and DOD programs, 163–164
turf disputes over control of, 32–33, 49, 188–189
Weinberger, Caspar, 17
White House Bureau of the Budget (BOB), 38–39
White House Domestic Policy Council, 86
Williamson, Ray, 4, 8, 25
Wilson, James, 18, 23–24, 194, 195
Woods Hole conference (July 1983), 60
World War II, remote sensing technology in, 30

Zraker, Charles, 144
Zraker Committee, 144, 175